Katalog zur Ausstellung/
Catalogue for the exhibition
„Carl Remigius Fresenius“
Museum Wiesbaden
23 Aug 2018 – 20 Jan 2019

Carl Remigius Fresenius
Wiesbadener Ehrenbürger/
Honorary citizen of Wiesbaden
28 Dez 1818 – 11 Jun 1897

Museum Wiesbaden

HOCHSCHULE
FRESENIUS
UNIVERSITY OF APPLIED SCIENCES

Inhalt Contents

Vorherige Seite: Quantitatives Labor in der sogenannten unteren Weinstation des Chemischen Laboratoriums um 1900.

Previous page: Quantitative chemistry lab in the wine department of the Fresenius Chemical Laboratory around 1900.

Vorwort

Dieses Büchlein ist, wie alle Bücher, mit Vorsicht zu genießen. Es versucht die Annäherung an einen Menschen, der vor 200 Jahren geboren wurde und vor 121 Jahren starb. Er lebte also in einer von der unseren sehr verschiedenen Welt. Es möchte die große Lebensleistung des Wiesbadener Ehrenbürgers Carl Remigius Fresenius in den Facetten

1. begabter, fleißiger und redlicher Mensch
2. akribischer und wirkmächtiger Wissenschaftler
3. dienstleistender und innovativer Problemlöser
4. hingebungsvoller und nachhaltig wirksamer Lehrer
5. einfalls- und erfolgreicher Unternehmer und
6. gemeinwohlorientierter Bürger

gegenwärtig werden lassen. Dabei kommt er hin und wieder selbst zu Wort – in privaten, wissenschaftlichen und poetischen Texten. Schriftstücke, Anekdoten, Urteile von Zeitgenossen und Nachgeborenen, Porträts, Abbildungen von Laborgeräten sind Mosaiksteine, die zusammen ein anschauliches Bild des als „Vater“ oder „Nestor“ der Analytischen Chemie gepriesenen Chemikers ergeben sollen.

Dabei soll nicht vergessen werden, dass sein Lebenswerk fortbesteht und reiche Früchte trägt. Deshalb dürfen aktuelle Bezüge nicht fehlen: Beispiele zum Stand der Wissenschaft und Praxis in den von ihm bearbeiteten Feldern und Kurzbeschreibungen der Zweige des großen Baums, den er mit der Gründung seines Laboratoriums 1848 gepflanzt hat.

Leo Gros

Foreword

Like all books, this one is to be read with caution. It is an attempt to come closer to a man who was born two hundred years ago and died considerably more than a century ago and thus lived in a world very different from ours today. Its aim is to pay tribute to the life achievement of Carl Remigius Fresenius, Freeman of the City of Wiesbaden, in the light of the following facets:

1st a gifted, hard-working and upright man
2nd a painstaking and highly influential scientist
3rd an innovative solver of problems in the service of others
4th a dedicated and enduringly influential teacher
5th a resourceful and successful entrepreneur
6th a citizen concerned for the common good

In this tribute he is also allowed to speak for himself – in private, scientific and poetic texts. These quotations are complemented by various writings, anecdotes, assessments on the part of contemporaries and non-contemporaries alike, portraits, and photos of laboratory instruments, which have been assembled like the stones of a mosaic to make a vivid picture of the chemist who has come to be known as the 'Father' or 'Nestor' of analytical chemistry.

The reader of this book should not forget that the life work of Carl Remigius Fresenius did not come to end with his death and that it is stil bearing rich fruit today. For this reason, it is only right that this book frequently refers to the present time with examples illustrating the state of knowledge and practice in his fields of activity and with short descriptions of the branches of the great tree that he planted through the foundation of his laboratory in 1848.

Leo Gros

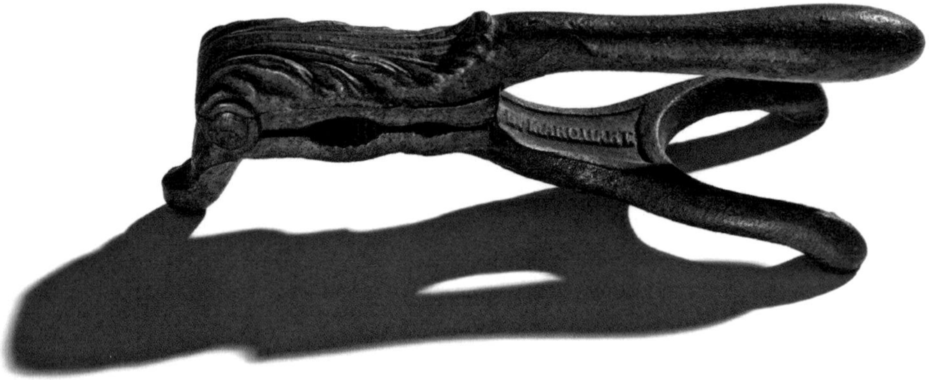

Handpresse zum Geschmeidigmachen von Korken mit Prägung der Geräte- und Chemikalienhandlung Marquardt, Bonn. Im Labor von Clamor Marquardt exeperimentierte Remigius Fresenius 1840.
Sammlung Friedrich

Handheld press for smoothing corks down, bearing a stamp of the Marquardt company that traded in Bonn lab equipment and chemicals. Fresenius did experimental work in Marquardt's laboratory in 1840. Collection Friedrich.

Der Mensch[1]

Wes Geistes Kind ist dieser „Remi"? Vom Werden eines großen Mannes

The Man[1]

What spirit shaped Remi's mind? The genesis of a great man

1

„Rrupp, ein anderes Bild" – so soll Remigius Fresenius gesagt haben, wenn er von einer seiner vielen Tätigkeiten in die andere wechselte. Das fiel ihm umso weniger schwer, weil er eine ausgesprochene Begabung für methodisches Vorgehen in allen Dingen hatte. Seine Freunde nannten ihn scherzhaft „System". Er konnte sich intensiv einer Sache widmen, nahm sich aber auch Zeit für Erholung in der Natur, im Theater und auf der Jagd. In seinen Briefen und in Gedichten verarbeitete er Erlebtes und Erfahrenes. Diszipliniert, aber ohne sich

'Rrupp, over to a different picture' – this is what Remigius Fresenius is said to have remarked when changing from one of his many activities to another. Changing to and fro was all the easier for him because in all spheres of life he had an exceptional talent for proceeding methodically. His friends jokingly called him 'System'. Even when he was intensely occupied with a certain sphere of activity, he would also find time for recreation in the world of nature, at the theatre, or going out shooting. He hetzen zu lassen, meisterte er die vielfältigen Aufgaben seines abwechslungsreichen Alltags. Seine Analysen waren sehr präzise, seine Problemlösungsstrategien einfallsreich. So wenig er in der Theorie der Wissenschaft steckenblieb, so sehr wusste er sein Forschen, Lehren und praktisches Arbeiten im Kontext seiner Wissenschaft und seiner Zeit einzuordnen.

Bei all dem war er optimistisch, ausdauernd, entschlussfreudig und seelisch ausgeglichen. So jedenfalls beschreiben ihn Freunde und Familienmitglieder. Wie bildet sich ein solcher Charakter, wes Geistes Kind ist Remigius?

Seine Eltern waren bereits 38 Jahre alt, als „Remi", ihr drittes und jüngstes Kind, ihr einziger Sohn, geboren wurde. Remi – so nannten ihn Familie und Freunde lebenslang.

Die oft kränkliche, bereits 1841 verstorbene Mutter und der als Anwalt beruflich sehr eingespannte Vater ließen ihn früh selbständig werden. Liest man den umfangreichen Briefwechsel mit seinem Vater Jacob Heinrich Samuel und anderen Familienmitgliedern, so zeigt sich dennoch eine starke familiäre Bindung. Den Rat seines lebensklugen, frommen Vaters holte er vielfach ein, traf jedoch zunehmend eigenständige, auch von der Meinung des Vaters abweichende Entscheidungen.

Offenbar genoss er eine anregungsreiche, naturnahe Kindheit und Jugend. Ein Brief aus dem Jahr 1852 mutet wie eine aus eigener Erfahrung gespeiste pädagogische Einsicht an: Als ein Frankfurter Lehrer den dann 33jährigen Mann fragt, ob er bereit sei, „Knaben" in Chemie zu unterrichten, entgegnet der, Kinder sollten ihre Jugend genießen. Man möge sie lehren, ihre Sinne zu gebrauchen, zu denken, und sich an

continued to mull over his experiences and discoveries in letters and poems. He showed great discipline, without ever hurrying, in mastering the diverse tasks of his everyday life. He was very precise in his analyses and inventive in thinking out strategies for solving problems. Far from getting bogged down in the theory of science, he put his research, his teaching and his practical work firmly into the context of his science and his time.

In all this he was, by all accounts, optimistic, persevering, decisive and psychologically well-balanced. How is a character such as his formed?

His parents were both thirty-eight years of age when Carl Remigius was born, their third and youngest child and their only son. 'Remi' was what his family and friends called him his whole life long. Remi was allowed a great measure of independence at an early age, both by his mother, who was often ill and died in 1841, and his father, who was a busy lawyer. However, if one reads Remi's extensive correspondence with his father Jacob Heinrich Samuel and other relatives, one soon notices that he had strong family ties. He often sought his wise and pious father's advice but increasingly also made his own decisions.

In his childhood and youth his life was clearly full of stimulation and close to the world of nature. A letter from the year 1852 reads like the expression of a pedagogical insight drawn from his own experience. When asked whether he would be willing to teach chemistry to boys, the reaction of the 33-year-old Fresenius was that the young should

Gründlichkeit in Wissen und Verstehen zu gewöhnen.

Zu all dem hatte Remi als Kind und Jugendlicher reichlich Gelegenheit. Die Musterschule in Frankfurt war inspiriert von Rousseau und Pestalozzi. Drei Jahre lernte er dann im Institut von Bender in Weinheim, das man heute einem Landschulheim mit humanistischem Gymnasium vergleichen könnte. Ausgedehnte Wanderungen und Exkursionen, Sport und Naturerleben gehörten zum Programm. Da war für Remi der Übergang in das später von seinem Sohn Wilhelm als „verknöchert" bezeichnete Frankfurter Gymnasium nicht leicht. Bereits in diesen jungen Jahren hatte sich das Interesse des Advokatensohns an Naturwissenschaften gefestigt. Seine Apothekenlehre bot die Möglichkeit zu praktischer Ausbildung und mannigfacher Anregung, forderte ihn aber auch voll und ganz. In seinen frühen Gedichten beklagt er

enjoy their youth. They should be taught to use their senses, to think, and to acquire the habit of thoroughness in knowledge and understanding.

As a boy, Remi had ample opportunity for all this. The Musterschule in Frankfurt where he went first was inspired by Rousseau and Pestalozzi. He then spent three years at Bender's in Weinheim, which was a kind of country boarding school offering a humanistic education. The overall timetable included extensive hikes and excursions, sport, and experience of the world of nature. As a result, it was not easy for Remi to make the transition to Frankfurt Grammar School ('Gymnasium'), which his son Wilhelm later called 'ossified'. Even in these early years the lawyer's son developed a strong interest in the natural sciences. His apprenticeship as a pharmacist, as well as giving him the opportunity

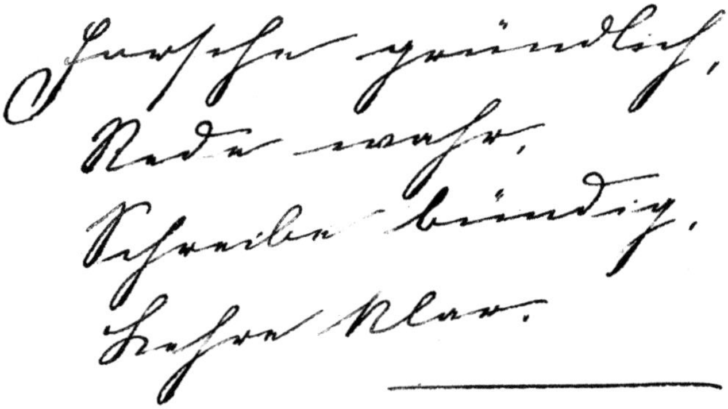

Handschriftliches Epigramm von Carl Remigius Fresenius. Aus einem Poesie-Album im Familienbesitz

Forsche gründlich,
Rede wahr,
Schreibe bündig,
Lehre klar.

Handwritten epigram of Carl Remigius Fresenius. From a family album.

Research: profound,
True be your speech,
Your writing: sound,
And: clearly teach.

den Mangel an Freiraum, den er sich gleichwohl verschafft: Im Gartenhaus seiner Familie in Bornheim experimentiert er eigenständig und hört Vorlesungen im Senckenberg'schen Institut und im Physikalischen Verein. Sein Studium in Bonn 1840/41 bei namhaften Professoren betreibt er gewissenhaft (seine Vorlesungsmitschriften sind uns erhalten geblieben). Prägend dürfte jedoch gewesen sein, dass er im Privatlabor von Clamor Marquart arbeiten konnte. Seine über 100 Analysen dort hat er knapp und genau protokolliert. In dieser Zeit entstand seine „Anleitung zur qualitativen Analyse". Eigentlich als Dokumentation seines Lernfortschritts gedacht, brachte er sie auf Drängen von Marquart als 82seitiges Büchlein heraus. Marquart riet ihm, bei Liebig in Gießen zu studieren. In außergewöhnlich kurzer Zeit wurde er dessen Assistent, promovierte und erhielt 1843 die Lehrbefugnis (venia legendi) – einer Professur vergleichbar. Als er seine „Antrittsvorlesung" (» Kapitel 2) hielt, war er wenig über 25 Jahre alt. 1845 nahm er den Ruf auf die Landwirtschaftsschule in Wiesbaden an und gründete 1848 sein eigenes Laboratorium (» Kapitel 5). Er wurde in den fast fünfzig Jahren seiner Tätigkeit im Laboratorium ein einflussreicher Lehrer von über 2300 Schülern, darunter namhaften Wissenschaftlern und Industriellen. (» Kapitel 4)

Die Auflagenzahl und die durch zahlreiche Übersetzungen belegte Reichweite seiner analytischen Lehrbücher und der Einfluss seiner heute noch erscheinenden Fachzeitschrift (» Kapitel 2.3) sind Legende. Sein Laboratorium war Bildungsstätte und zugleich ein weithin geschätztes Dienstleistungsunternehmen für ein breites

for a practical training and stimulating his mind in many ways, also made far-reaching demands upon him.

In his early poems we find him lamenting his lack of space for free activity, but at the same time he created it for himself, conducting his own experiments at his family's garden house in Bornheim and attending lectures at the Senckenberg Research Institute and at Frankfurt's 'Society of Physics'. In Bonn he worked hard as a student under well-known professors (we still have his lecture notes). A more formative factor, however, is very likely that he was able to work in the private laboratory of Clamor Marquart, where he made concise and accurate records of over a hundred analyses. It was during this time that he wrote his 'Introduction to Qualitative Analysis'. Although it was in fact only intended as a documentation of his progress as a student in this field, Marquart urged him to have it published, which he did, in the form of a small book 82 pages long. On Marquart's advice he studied in Gießen under Liebig, whose assistant he soon became. He took his doctorate and in 1843 received his licence to teach ('venia legendi'), comparable with a professorship, giving his inaugural lecture (» chapter 2) at the age of only 25. In 1845 he accepted an appointment at the Agricultural Institute in Wiesbaden and in 1848 founded his own laboratory (» chapter 5). In the course of a career lasting almost fifty years he was the influential teacher of over 2,300 students, a number of whom were to make names for themselves as scientists and industrialists (» chapter 4).

Szene aus dem Liebigschen Labor, möglicherweise um 1840

Scene in the Liebig Laboratory, possibly around 1840

Der Künstler Hermann Schies in seinem Atelier bei der Arbeit an der Büste von Carl Remigius Fresenius. Ein Gipsabdruck im Besitz des Museums Wiesbaden ist in der Ausstellung zu sehen. Die Marmorbüste steht im Traditionsraum der Hochschule Fresenius in Idstein.

The artist Hermann Schies in his studio working the bust of Carl Remigius Fresenius. The Landesmuseum Wiesbaden has a plaster cast which is shown in the exhibition. The bust is displayed at the Hochschule Fresenius in Idstein.

Von eigener Hand (Autograph)

In einem Brief an den Vater vom 19.06.1843 spricht Carl Remigius Fresenius von seinem „Plan, Autorität in diesem Felde [der analytischen Chemie] zu werden, wozu ich auf dem besten Wege bin“.

From Remi’s hand (autograph manuscript)

In a letter to his father of 19 June 1843 Carl Remigius Fresenius writes of his ‘plan to become an authority in this field [analytical chemistry], which I am on a very good path to fulfilling’.

Spektrum von praktischen Fragestellungen (» Kapitel 3).

Als Remigius Fresenius 1897 knapp achtzigjährig starb, hinterließ er seinen beiden im Laboratorium arbeitenden Söhnen und seinen Mitarbeitern ein wohl bestelltes Feld, seiner Frau umfangreiche Aktienpakete und Immobilien. Er war eben auch ein erfolgreicher Unternehmer (» Kapitel 5) und ein geschätzter Mitbürger (» Kapitel 6).

Fresenius war nicht zuletzt ein „Familienmensch" – liebender Gatte seiner beiden Frauen, und ein zugewandter Vater. Aus seiner ersten Ehe überlebten ihn 7 Kinder. Für jedes seiner zahlreichen Enkelkinder gab es einen eigenen „Großelterntag" am Tisch der Alten.[2]

In der Nacht vom 10. auf den 11. Juni 1897 stirbt Fresenius nach kurzem „Unwohlsein", mit knapp 80 Jahren. ●

His analytical textbooks are legendary for the number of editions they went into and the scope of their influence, which is demonstrated not least by the number of translations; similarly legendary is the influence of the specialist journal he founded, which is still published today (» chapter 2.3). And as well as being a place of training and education, his laboratory also functioned as a widely esteemed service enterprise for a wide range of practical matters (» chapter 3).

When Remigius Fresenius died in 1897 in the eightieth year of his life he bequeathed a well-ordered and flourishing heritage to his sons, both of whom worked at the laboratory. His widow's inheritance also contained a wide range of share packages and items of real estate, for he was a successful man of business (» chapter 5). Furthermore, he was a highly regarded member of society (» chapter 6).

In addition, and not least, Fresenius was a 'family man' – a loving husband of two wives and an interested and caring father. Seven children survived him from his first marriage. Each one of his numerous grandchildren was allotted a special 'grandparents' day' with a celebration around his family dining table.[2]

In the night of 10/11 June 1897, after having felt 'unwell' for a short period, Fresenius died at the age of almost eighty. ●

Schwefelsäure vorsichtig in den Probircylinder einfliessen, so dass zwei Schichten entstehen. Ist Salpetersäure zugegen, so färbt sich die Berührungszone braun, bei sehr geringen Mengen gelb.

Peter T. Austen und J. Chester Chamberlain*) empfehlen zum Nachweis der Salpetersäure statt Eisenvitriol schwefelsaures Eisenoxydul-Ammon zu benutzen, da letzteres Salz in Lösung vorräthig gehalten werden könne.

Ein neues nichtmetallisches Element hat Clemens Winkler**) entdeckt und demselben den Namen Germanium gegeben. Er theilt darüber Folgendes

»Im Sommer
grube« bei Freib
sehen, in welchem
die er »Argyro
einer vorläufigen
Hauptbestandtheile
das Vorhandensein
fallend und inter
Erzgängen bisher

Bei der von
sich, dass der ge
trägt; ausserdem
gewandten Materi
sowie sehr gering
So oft und so sor
mochte, immer s
Verluste ab, ohn
Untersuchung mög
decken.

Nach mehrw
Bestimmtheit auss
sehr ähnliches, a
enthält, welchem

*) American c
**) Ber. d. deut
gesandt.

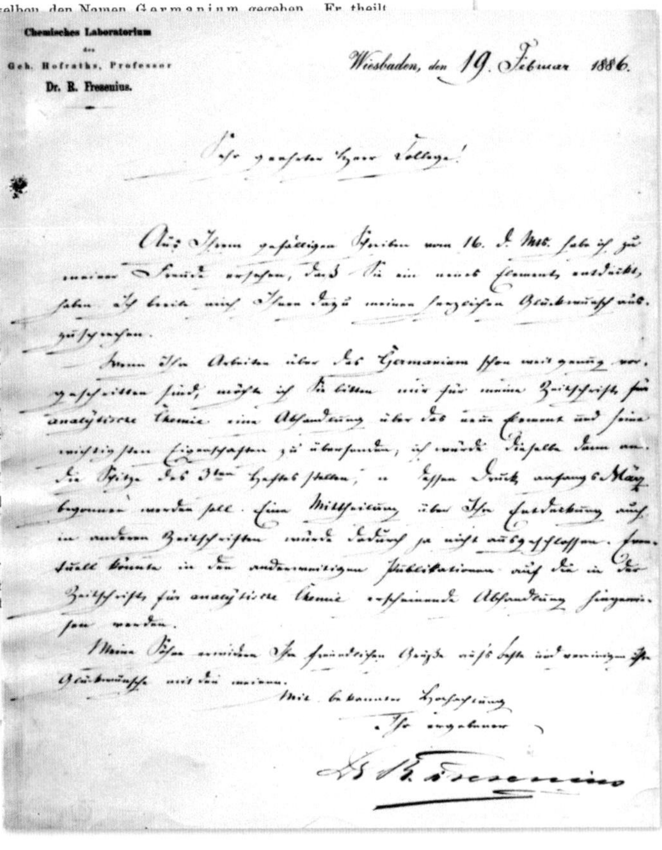
Chemisches Laboratorium des Geh. Hofraths, Professor Dr. R. Fresenius.

Wiesbaden, den 19. Februar 1886.

Brief von C.R. Fresenius an Professor Clemens Winkler, Freiberg, vom 19.02.1886. Kurz zuvor hatte Winkler in einem Erz ein bislang unbekanntes Element gefunden, das er Germanium nannte. Fresenius bittet um einen Beitrag über Germanium und seine Eigenschaften für seine Zeitschrift für Analytische Chemie. Links daneben die umgehend in der nächsten Nummer der Zeitschrift publizierte Notiz.

Letter C.R. Fresenius wrote to Professor Clemens Winkler, Freiberg on 19 February, 1886, shortly after Winkler had discovered a hitherto unknown element which he named Germanium. Fresenius asks for a contribution about Germanium and its properties for his journal. Left: the short notice published in the subsequent issue of the journal.

Der Wissenschaftler

„Wie ein Fuß, der die fremde Küste betritt ...“[3]

The Scientist

‘Like a foot that steps upon a foreign shore ...’[3]

2

Ein Gymnasiallehrer, so berichtet Remigius seinen Kindern, habe seinen Entschluss eine Apothekenlehre zu beginnen mit den Worten kommentiert: „So, Sie wollen die Wissenschaft an den Nagel hängen, das ist auch besser so.“[4] Wäre „Remi“ wirklich dem Rat seines Lehrers gefolgt, gäbe es dieses Büchlein nicht.

Fresenius wurde allerdings kein abgehobener Theoretiker. Eine Wissenschaft „im Elfenbeinturm“ war seine Sache nicht. Gut, dass ein Brief seines Vaters vom 2. März 1842 zur Frage der Berufswahl

When Remigius decided to embark upon an apprenticeship as a pharmacist, so he told his children, one of his grammar school teachers made the following comment: ‘Aha, so you want to give up science? Well, that’s a good thing.’[4] Had ‘Remi’ followed his teacher’s advice, this little book would never have been written.

However, Fresenius did not become a theoretician with his head in the clouds. ‘Ivory-tower’ science was not his thing at all. It is good that a letter from his father erhalten geblieben ist. Nachdem er vorausgeschickt hat, dass er sich „Stillschweigen“ auferlegt in einer so lebensentscheidenden Frage, rückt er doch mit der Erfahrung heraus, „daß hervorragende Geister, welche die Gränzen der Wissenschaft vorrücken u. ganz neue Bahnen brechen, selten sind und den Lohn ihrer Anstrengung hauptsächlich in sich selbst und in der Wissenschaft suchen u. finden müssen“. Beschreibt er damit nicht das, was wir heute Grundlagenforschung nennen? Nach seiner Beobachtung ergeht es anwendungsorientierten Wissenschaftlern besser, da sie „in der Regel eher auf ein emsiges, glückliches Leben u. auf den Dank und Lohn der Mitwelt rechnen können. Es ist der Unterschied zwischen Genie und Talent; (Kometen und Sonnen.)“

Sein Vortrag zur Erlangung der Professur, dessen Manuskript uns erhalten blieb, ist kein Chemiker-Kauderwelsch. Er liest sich wie eine Kombination aus naturphilosophisch-wissenschaftsgeschichtlicher Einführung und hochschulpolitischen sowie chemiedidaktischen Überlegungen. Er enthält bereits wie Leitmotive die selbst gewählten Zielsetzungen seiner Lehre und Forschung: Erweiterung des Erfahrungs- und Erkenntnisraums durch wohlüberlegte Versuche, Herausarbeiten der Logik die sich dabei ergibt und strikte Vermeidung von, so würden wir es heute ausdrücken, einseitigem Fachidiotentum zugunsten einer ganzheitlichen Betrachtung.

Als Remigius Fresenius sich 1845 entschließt, den Ruf des Herzogs von Nassau an das Landwirtschaftliche Institut Hof Geisberg in Wiesbaden anzunehmen und 1848 sein eigenes Laboratorium

of 2 March 1842 concerning his choice of career has been preserved. After starting off by writing that the importance of the matter has caused him to put a seal upon his lips, he nevertheless comes out with advice from his experience, namely, ‘that outstanding minds who push back the bounds of science and break entirely new ground are rare, and that they have to seek and find the reward for their efforts principally within themselves and in science.’ With these words, was he not describing what we today know as ‘fundamental research’? According to his observation, scientists who are oriented towards applications have a better time of it because they ‘as a rule can count on having a busy and happy life and on receiving the gratitude and rewards of society. It is the difference between genius and talent (comets and suns).’

The lecture Fresenius gave to inaugurate his professorship, the manuscript of which has been preserved, is far from being couched in chemists’ jargon. It reads like a combination between an introduction to natural philosophy and history of science and a collection of thoughts on university politics and chemical didactics. Furthermore, it already contains, as leitmotifs, the goals that he had chosen for his teaching and research: the extension of experience and knowledge through well-considered experiments, the working out of their logical consequences, and the strict avoidance of what we would today call one-sided nerddom in favour of the cultivation of a holistic approach.

In 1845 Remigius Fresenius accepted the invitation of the Duke of Nassau to

Fresenius Will Bullock Gardener Hofmann
im Liebigschen Laboratorium in Gießen nach einer Daguerrotypie
anfangs der vierziger Jahre.

Fresenius im Kreis seiner Gießener Mitstudenten Will und Hofmann und zweier englischer Besucher des Liebig'schen Laboratoriums. Sie sollten im Auftrag der Royal Society of Chemistry (London) einen Kandidaten für eine Professur in London aussuchen. Sie konnten sich nicht entscheiden und schickten ein Foto nach London.[5] Wäre nicht Hofmann, sondern Fresenius ausgewählt worden, gäbe es diesen Katalog vielleicht nicht. Bullock wurde übrigens Fresenius' erster englischer Übersetzer.

Fresenius with his fellow students Will and Hofmann and two English visitors to Liebig's laboratory. They came with the task of choosing a candidate for a chair of Analytical Chemistry in London. They did not come to an agreement and sent a photo to London.[5] Had the final decision been in favour of Fresenius instead of Hofmann, this catalogue perhaps would not have been printed. Bullock, it may be noted, was to become Fresenius' first English translator.

gründet, hat er sich offenbar entschieden, Talent statt Genie und lange wärmende Sonne statt aufstrahlender und dann langsam verglühender Komet zu sein. Sein Laboratorium ist eine Stätte der angewandten Wissenschaft, in der er durchaus, mit seines Vaters Worten gesagt, „neueste Entdeckungen" im Bereich der analytischen Chemie macht, vorzüglich aber auch solche prüft und „ins Leben einführt". Wir würden es heute so ausdrücken: Er verschreibt sich ganz und gar der angewandten Forschung und der praktischen Anwendung sowie der Ausbildung von Menschen, die das Erlernte in Zukunft, in ihrem Lebenskreis und in vielen Ländern der Erde praktizieren und weiter verbreiten werden. Dazu hat er zunächst mit seinem Erstlingswerk „Anleitung zur qualitativen Analyse" die Ergebnisse der Wissenschaft „gemeinfaßlich" dargestellt. Dieses Buch so wie die bald darauf herausgebrachte „Anleitung zur quantitativen Analyse" wurden zu vielfach aufgelegten und übersetzten Standardwerken bis in die erste Hälfte des 20. Jahrhunderts.

1862 gründet er seine eigene Fachzeitschrift, deren Schriftleitung er lebenslang innehat und die in europäischem Gewand heute noch erscheint (» Kapitel 2.3).

Über 450 Publikationen unter seinem Namen sind bekannt. Schon 1843 reiste er zur Versammlung deutscher Naturforscher und Ärzte nach Graz. Als sie in den Jahren 1852, 1873 und 1887 in Wiesbaden tagte, war er ein geschätzter Geschäftsführer und Gastgeber dieser wissenschaftlichen Großereignisse.

Mit der Wahl seines Berufs als Forscher, Lehrer und Dienstleister, der ihm Berufung war, hat er wohl richtig gelegen

take up a post at the Hof Geisberg Agricultural Institute in Wiesbaden, subsequently founding his own laboratory in 1848. This makes it clear that he had decided to give talent priority over genius, and to be a sun rather than a comet, that is, to strive to emit warmth for a long period of time rather than to briefly shine brightly and then slowly burn out. His laboratory was a place dedicated to applied science. Although, to use his father's words, he indeed made 'the newest discoveries' in the field of Analytical Chemistry there, his principal goal was to test such discoveries and to 'introduce them into life'. Today we would put it differently, namely, that he committed himself fully and wholly to pursuing applied research and practical applications, and to training students who would subsequently apply and propagate what they had learnt in their own sphere of life and in many countries of the world. This was why he started by presenting his findings 'in readily comprehensible form' in his first work, the Introduction to Qualitative Analysis and the Introduction to Quantitative Analysis that followed soon after. Both works went into many editions and were translated into many languages, and remained standard works until well into the first half of the twentieth century. In 1862 he founded his own specialist journal, which he edited throughout his life and which lives on today as part of a Europe-wide periodical (» chapter 2.3).

Over 450 publications are known to have appeared under his name. As early as 1843 he travelled to Graz to participate in the Assembly of German Natural und nach einem „emsigen, glücklichen Leben" den reichen Dank seiner Mit- und Nachwelt dafür erhalten.

Er wäre hoffentlich zufrieden zu sehen, wie die Forschungsinstitute für Analytik (IFAR) und Bioanalytik (IBR) an der Hochschule Fresenius mit modernsten Geräten und in Zusammenarbeit mit vielen Partnern in Wissenschaft und Praxis die Probleme unserer Zeit bearbeiten und dabei die Studierenden der Hochschule Fresenius einbeziehen. ●

Scientists and Physicians. When the conference took place in Wiesbaden in 1852, 1873 and 1887, he earned widespread esteem for his management and hosting of this major scientific event.

He was certainly right in his choice of career. Indeed, the combination of researcher, teacher and service-provider was more a vocation than a career. It not only gave him 'a busy and happy life' but also earned him the gratitude of his contemporaries and of posterity.

He would hopefully have been happy to see how the Institute for Analytical Research (IFAR) and the Institute for Biomolecular Research (IBR) at the Fresenius University of Applied Sciences work on the problems of our time, using the most modern instruments, collaborating with many partners from research and enterprise, and involving Fresenius University students in their work. ●

Merk-würdig[6]

Liebig verlangt von Fresenius, einen Engländer in dessen Sprache zu unterrichten. Dessen Einwand, er sei des Englischen nicht ausreichend mächtig, lässt der Altmeister und Doktorvater nicht gelten: „Das muss man können, darauf kann ich keine Rücksicht nehmen." Fresenius schickt sich drein und tut sein Bestes. Beim Versuch einer englischen Version von „eine Lösung zur Trockene eindampfen" spricht Fresenius in Anlehnung an das ihm vertraute Latein von "evaporate the solution to siccness" statt von „dryness". Ein herrlicher „Latinismus", der den Engländer stutzen macht – bedeutet doch „sick" krank oder unwohl!

Note-worthy[6]

On one occasion, Liebig instructed Fresenius to teach an English student in his own language. When Fresenius objected on the grounds of his insufficient mastery of English, his master and doctoral supervisor rejected his argument: 'I cannot take that into consideration – this is simply something one must be able to do.' Fresenius threw himself into the task and did his best. However, when attempting to translate 'eine Lösung zur Trocken eindampfen', he fell back on his knowledge of Latin and came up with 'evaporate the solution to siccness' instead of 'dryness'. A wonderful Latinism that understandably made the Englishman stop in his tracks because it sounded like 'sickness'!

2.1

Die Qualitative Analyse: Was ist drin?

„... die Sprache zu erlernen“[7]

Qualitative analysis: What's in it?

'... to learn the language'[7]

Wir alle wissen: Kochsalz löst sich in Wasser, Weinstein nicht, Kalk löst sich in Säure, Lauge greift Glas an. Die Eigenschaftsunterschiede von Stoffen, vor allem ihre unterschiedlichen Löslichkeiten in Anwesenheit des Gases Schwefelwasserstoff, sind der Schlüssel zu Trennverfahren, die es im 19. Jahrhundert erstmals erlaubten, Gemische systematisch auf ihre Bestandteile zu untersuchen. Was wohl Remis Eltern gesagt haben mögen, als ihr Sohn mit noch nicht 18 Jahren im Bornheimer Gartenhäuschen mit jenem Gas experimentierte, das nach faulen Eiern riecht und in „Stinkbomben“ wissenschaftsfremden Zwecken dient?

Im Bonner Labor von Clamor Marquart vertieft Fresenius diese Kenntnisse. Dabei fußt er auf der Anleitung dieses erfahrenen Apothekers und kennt die Werke des

We all know that table salt dissolves in water but tartar does not, that lime dissolves in acid, and that an alkaline solution corrodes glass. The differences in properties between substances, especially their various levels of solubility in the presence of hydrogen sulphide, are the key to separation procedures that in the nineteenth century first made it possible to examine the components of composite substances. What will Remi's parents have said when their son, not yet eighteen years of age, started experimenting in their garden house at Bornheim with the gas that smells of bad eggs and is used for non-scientific applications in stink bombs?

In Clamor Marquart's laboratory in Bonn, Fresenius deepened the knowledge

Carb.
Spar.
Calc.

Lötrohr mit Holzkasten und Reagenzbehältern im hinteren Teil. Es dient zur qualitativen Erkennung von Stoffen z.B. in Erzproben. Die Funktion des Lötrohrs wird in einem Videofilm an einem Beispiel in der Ausstellung gezeigt. Um 1913. Sammlung Friedrich.

Blow pipe with wooden box and reagent holders in the rear part. It served for the qualitative detection of elements, in e.g. minerals. Its function is shown in a video in the present exhibition. Around 1913. Collection Friedrich.

großen schwedischen Chemikers Berzelius und des Berliner Professors Heinrich Rose. Fresenius' 1841 erschienene erste Auflage der „Qualitativen Analyse" erklärt Schritt für Schritt und streng systematisch die praktische Arbeitsweise. Das Buch erlebt bereits in den ersten 20 Jahren zehn Auflagen und wird mehrfach ins Englische, Französische, Spanische, sowie ins Italienische, Russische, Niederländische, Ungarische, Japanische und Chinesische übersetzt. Fresenius schaut seinen Schülern bei der Arbeit zu, erfasst Unklarheiten und Probleme und setzt diese Erkenntnisse in didaktisch bessere Texte um. Schrittweise fügt er spezielle Informationen zur Untersuchung von Wässern, Lebensmitteln und vielen technischen Produkten ein. So werden aus den anfangs 81 Seiten schließlich fast 900.

Was ist drin? Auf diese Frage antwortet die qualitative Analyse. Sie ist bis heute, auch an der Hochschule Fresenius, eine hervorragende Schule für den Umgang mit Stoffen und sauberes Arbeiten. Viele Alltagsbeobachtungen fasst die Chemie in wenigen, allgemein gültigen Regeln zusammen. Wer die „Quali" verstanden hat, kann das Sprudeln einer Schorle, die Wirkungsweise von Backpulver oder das Entkalken einer Kaffeemaschine mit wenigen einfachen Kurzformeln, genannt Gleichungen, beschreiben und verstehen. ●

he had acquired at Bornheim, basing his studies on the instruction given him by the experienced apothecary Marquart and knowing the works of the great Swedish chemist Berzelius and Professor Heinrich Rose of Berlin. The first edition of Fresenius's textbook 'Qualitative Analysis' of 1841 gives a strictly systematical explanation, step-by-step, of the practical working procedure. In the following 20 years the book went into ten editions, with several editions in English, French and Spanish, and individual editions being translated into Italian, Russian, Dutch, Hungarian, Japanese and Chinese. Through observation of his pupils at work, Fresenius was able to take note of unclear points and problems and compose texts of higher didactic quality. Gradually he added special information regarding the investigation/analysis of waters, foods and many technical products. As a result, the 82 pages of the first edition finally grew to almost nine hundred.

What's in it? This is the question that is answered by qualitative analysis. In our own day too, not least at the Fresenius University of Applied Sciences, qualitative analysis remains an excellent school for correct and precise work with substances. In chemistry, many observations from everyday life are rationalized in a small number of generally valid rules. Students who have understood 'qual' can describe and understand, by means of a few simple formulae known as equations, all sorts of phenomena from the fizz in a spritzer or the effect of baking powder in dough through to descaling a coffee machine. ●

Reagenzgläser (um 1930) in einem hölzernen Reagenzglasständer (um 1900). Sie dienen als Glasgefässe für Versuche im Kleinmaßstab, z.B. in der qualitativen Analyse. Sammlung Friedrich

Test tubes (around 1930) in a wooden rack (around 1900). They are used for small scale reactions (e.g. in qualitative analysis). Collection Friedrich

Bunsenbrenner. Unentbehrliches Gerät zur Erzeugung heißer Flammen. Viele Reaktionen laufen nur in der Hitze. Zwischen 1880 und 1910. Sammlung Friedrich.

Bunsen burner. Again an indispensable piece of equipment for producing hot flames. Many chemical reactions require high temperatures. Between 1880 and 1910. Collection Friedrich.

Kipp'scher Apparat zur Erzeugung des in der qualitativen Analyse als Reagenz dienenden Gases Schwefelwasserstoff (H_2S). ca. 1915. Sammlung Friedrich

Kipp apparatus for hydrogen sulphide (H_2S) production. Around 1915. Collection Friedrich

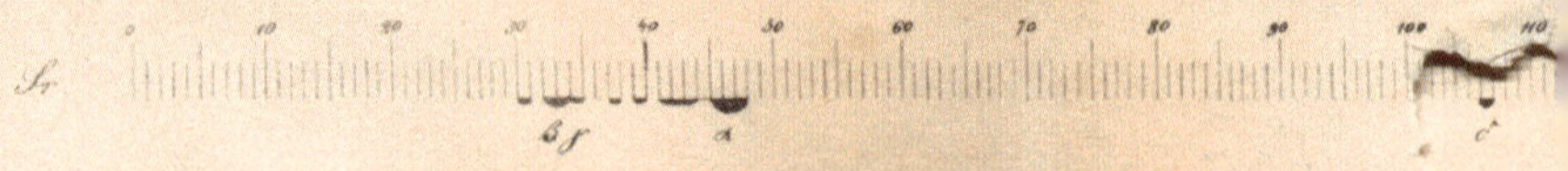
Sr
0 10 20 30 40 50 60 70 80 90 100 110
β γ
α
δ

Fig. 2

A a B C D E b F
Cs.
Rb.
Ka
Na
Li
Sr
Ca
Ba

Fig. 1

Fig. 29 a.

Handkolorierter Druck der von Bunsen und Kirchhoff mit ihrem Gerät aufgenommenen Spektren aus dem Anhang zur ersten Ausgabe der Fresenius' Zeitschrift für Analytische Chemie 1862. Rechts oben Abbildung des Spektroskops aus Fresenius' Lehrbuch der Qualitativen Analyse von 1862.

Hand-coloured print of the spectra Bunsen and Kirchhoff had measured with their apparatus. From the supplement to the first issue of Fresenius' Zeitschrift für Analytische Chemie 1862. Above, to the right, engraving of the spectroscope taken from Fresenius' textbook about qualitative analysis 1862.

Worte des Meisters

„Begleiten wir, um uns die Sache deutlicher zu machen, einen jungen Mann ... bei dem wir nur theoretische Kenntnisse in der Chemie und ihren Hülfswissenschaften voraussetzen. Er kann auf diesem seinem Standpunkte mit der Natur noch nicht sprechen, wenn ich so sagen darf; sein Zweck muss vorerst seyn, die Sprache, in der es allein möglich ist, zu erlernen, damit er alsdann selbstredend die Natur um ihre Wahrheiten befragen, selbstforschend derselben ihre Geheimnisse zu entlocken vermag. Die Sprache des Chemikers aber ist die der Experimente, ihre Laute sind die einzelnen Reaktionen; zu ihrer Erlernung wird jetzt allgemein die qualitative Analyse als der beste und kürzeste Weg betrachtet.“[7]

Words of the Master

'In order to make the matter clearer, let us consider a young man ... whom we shall assume to have only theoretical knowledge of chemistry and its auxiliary sciences. In his present position he cannot yet, if I may put it thus, speak with Nature; initially, his goal must be to learn the language in which this is alone possible, so that he himself may be able to ask Nature questions concerning her truths and through research may be able to coax her mysteries out of her. But the language of the chemist is that of experiments, and its sounds are those of the individual reactions; and it is now generally considered that the best and quickest way to learn this language lies in qualitative analysis.'[7]

2.2

Die Quantitative Analyse: Wieviel ist drin?

„Mit dem Wissen muss das Können sich vereinigen ...“[8]

Wer wie „Remi“ vorhat, eine Autorität auf dem Gebiet der Analytik zu werden, muss zur qualitativen ihre hohe Schule, die quantitative Analytik fügen. 1845 veröffentlicht er die erste, 518 Seiten umfassende Auflage. Die sechste erscheint 1875 bis 1887 und bringt es in zwei Bänden auf zusammen 1539 Seiten.

Wiederum systematisch führt der Autor in das Handwerk (die Operationen) ein. Er beschreibt die Stoffe, mit denen man den Gehalt einer Probe am zu untersuchenden Stoff bestimmt (die Reagenzien). Zwei Hauptmethoden sind die Titration oder Maßanalyse und die Gravimetrie oder Gewichtsanalyse. Ein Beispiel soll das erläutern. Die Titration des Essigsäuregehalts in Haushaltsessig macht man auch heute noch so, dass man die Probe

The Quantitative analysis: How much is in it?

‘Knowledge must be combined with skill ...’[8]

When someone intends, as ‘Remi’ did, to become an authority in the analytical field, they must complement qualitative analysis with its ‘haute école’: quantitative analysis. In 1845 Fresenius published the first edition of his work on quantitative analysis, which ran to 518 pages. By the time the sixth edition appeared in two volumes between 1875 and 1887, it had expanded to 1539 pages in all.

Systematically, as ever, the author introduces his readers to the craft of quantitative anlaysis (the operations). He describes the substances with the help of which the content of a given sample can be determined (the reagents). Two principal methods are titration (volumetric analysis) and gravimetry (analysis by weight). The former method can be

Spritzflasche mit Handhabe aus Holz (zum Anfassen bei Verwendung heißer Flüssigkeiten). Sie diient zum Beispiel dazu, ausgefällte Feststoffe mit heißem Wasser auf einem Filter auszuwaschen und so von anhaftenden löslichen Stoffen zu reinigen.

Wash bottle with wooden handle (to be used when manipulating hot liquids). It serves, e.g., for washing precipitated solids on a filter in order to remove adhering soluble substances.Collection Friedrich.

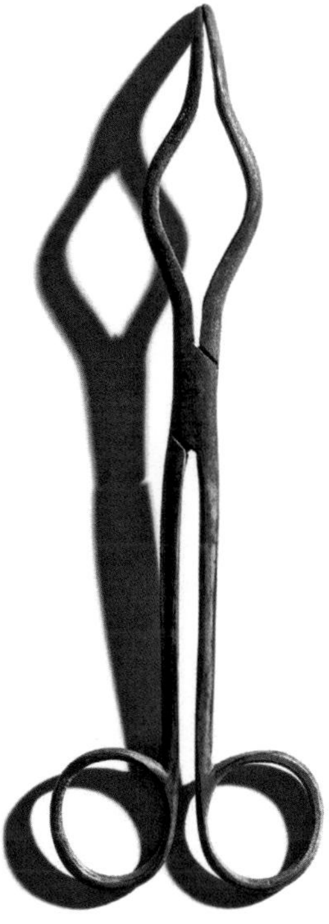

Tiegelzange zum Ergreifen, Festhalten und Manipulieren heißer Porzellantiegel. In solchen Tiegeln werden in der quantiativen Analyse zur Wägung gefällte Stoffe bis zur Gewichtskonstanz geglüht und anschließend im Exsiccator (» Bild S. 45) abgekühlt. Vernickelter Messingguß, um 1900. Sammlung Friedrich

Crucible tongs for seizing, holding and manipulating hot porcelain crucibles. In quantitative analysis, precipitates are ignited in these vessels until their weight is constant and then cooled down in a desiccator (» photo p. 45) Nickel plated brass cast, around 1900. Collection Friedrich.

tropfenweise so lange mit einer Lauge genau bekannten Gehalts versetzt, bis ein Farbindikator die vollständige Umsetzung der Säure anzeigt. Der Verbrauch an dem Reagenz Lauge erlaubt die Berechnung des Essigsäuregehalts.

Fresenius' Lehrbücher zur „Quanti“ gehen dann in einem zweiten Teil auf die Analyse der wichtigsten Untersuchungsgegenstände wie Wasser, Boden, Luft, Dünger, Pflanzen, Mineralien und andere technische Produkte ein. Zusammen mit der „Quali“ bildeten sie für sie viele Jahrzehnte lang weltweit die „Bibel der Analytik“.[9]

Auch wenn heute komplexe Analysengeräte solche Messungen automatisiert und in kürzerer Zeit sehr präzise erledigen, bleiben die guten alten Fertigkeiten unverzichtbar: Wenn wir die Proben nicht wie von Fresenius beschrieben richtig vorbereiten, misst das Gerät unser Ungeschick, nicht den Gehalt in der Probe. Dies gilt umso mehr, als moderne Analytik die Grenzen des eben noch Messbaren um viele Größenordnungen nach unten verschoben haben. ●

illustrated with an example. The titration of the acetic acid content in household vinegar is still carried out by adding to the sample drop-by-drop with an alkaline solution of a given known content until a colour indicates that the acid has been entirely converted. The acetic acid content can then be calculated on the basis of the quantity of the alkaline reagent used up.

Fresenius' textbooks on 'quanti' then proceed in a second part to consider the analysis of the most important objects of investigation such as water, soil, air, manure, plants, minerals and certain technical products. For many decades his textbooks on 'quali' and 'quanti' together constituted the 'Bible of chemical analysis'.[9]

Even though today's complex analytical instruments can carry out measurements such as these automatically and in a shorter time, the good old skills remain indispensable. If we do not prepare the samples correctly, in accordance with Fresenius' descriptions, the instrument does not measure the content in the sample but, rather, the incorrectness of our preparation. This is all the more true in that modern analysis has lowered the limits of the measurable by many orders of magnitude. ●

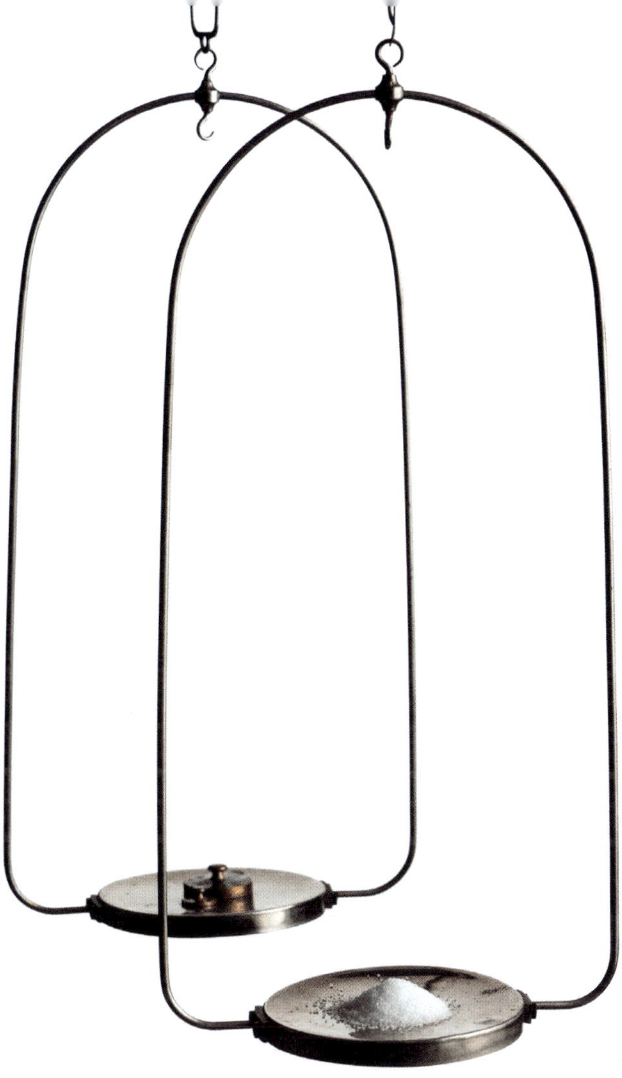

Detail einer Analysenwaage in einem Mahagonikasten, 1870, Ditmar & Vierth. Die Waage ist ein unentbehrliches und zentrales Instrument in der quantitativen Analyse. Wägen heißt in der Chemie soviel wie Moleküle zählen. Alltagsvergleich: Wer Schrauben wiegt, 220 g ermittelt und weiß dass eine Schraube 2 g wiegt, kann ausrechnen, dass er 110 Schrauben gewogen hat (ohne sie zu zählen). Sammlung Friedrich.

Detail of an a balance in a mahogany cabinet , 1870, Ditmar & Vierth. The balance is the indispensable and central instrument. In chemistry, weighing means counting molecules. An everyday example: If we weigh screws and find 220 g, knowing that one screw weighs 2 g, we know that there are 110 screws. Collection Friedrich.

Gewichtssatz fein Sartorius 100g–1g in 9 Gewichten und 14 Feingewichte unter Glas mit Pinzette in Mahagonikasten. Um 1870. Sammlung Friedrich

Set of balance weights Sartorius 100g - 1g in 9 weight pieces, 14 fine weight pieces under glass with forceps, in a mahogany case. Collection Friedrich.

Bürette für die Maßanalyse (Titration). Aus diesem mit Milliliter-Einteilung versehenenen Glasrohr werden in der quantitativen Analyse definierte Volumenmengen einer Reagenzlösung (z.B. Natronlauge) von bekanntem Reagenzgehlt mit einer in einem Erlenmeyer-Kolben vorgelegten Probelösung unbekannten Gehalts (z.B. Essigsäure in Haushaltsessig, Säuren in Wein) so lange versetzt, bis ein Farbumschlag den vollständigen Umsatz anzeigt. Der Flüssigkeitsablauf aus der Bürette wurde über einen Quetschhahn an einem Gummischlauch reguliert (heute dient dazu ein Glashahn). Die Titration wird heute meist automatisch durchgeführt. Geräte: Sammlung Friedrich

Buret for volumetric analysis (titration) and Erlenmeyer flask. From the glass tube calibrated in 0,1 milliliter steps defined volumes of a reagent (e.g. a sodium hydroxide solution) of known reagent concentration are added dropwise to a sample solution of unknown concentration (e.g. of household vinegar or acids in wine) until a colour change indicates full conversion of the acid to its salt. The addition of the reagent solution was controlled with a rubber tube (today, a glass stopcock is used). Most titrations are done automatically these days. Equipment: Collection Friedrich.

Exsiccator nach Fresenius. In diesem verschließbaren Glasgefäß trocknet man in der quantativen Analyse Feststoffe, damit sie wasserfrei gewogen werden können. Um 1880. Sammlung Friedrich.

Desiccator constructed by Fresenius. In this tightly closed glass vessel, substances are dried in quantitative analysis, before they are weighed. Around 1880. Collection Friedrich.

Worte des Meisters

(vom Wissenschaftshistoriker William H. Brock
auch „Hippokratischer Eid für Analytiker" genannt)[8]

„Jeder , der sich einigermaassen mit quantitativen Analysen beschäftigt hat weiß, dass sich, besonders am Anfange, zuweilen Fälle ereignen, in denen man Zweifel hegt, ob das Resultat genau ausfallen wird, oder in denen man gewiss ist, dass es nicht sehr genau ausfallen kann. Bald ist ein wenig verschüttet worden, ... bald zweifelt man, ob man sich im Wägen nicht geirrt habe, - bald stimmen zwei Analysen nicht recht überein. In solchen Fällen handelt es sich darum, dass man die Gewissenhaftigkeit habe, die Arbeit alsobald noch einmal zu machen. Wer diese Selbstüberwindung nicht hat, wer Mühe scheut, wo es sich um Wahrheit handelt, wer sich auf Schätzen und Muthmaassen einlässt, wo es die Erlangung positiver Gewissheit gilt, dem müssen Fähigkeit und Beruf zur Ausführung quantitativer Analysen ebenso gut abgesprochen werden, als wenn es ihm an Kenntnissen oder Geschicklichkeit gebräche. Wer seinen Arbeiten nicht selbst volles Vertrauen schenken, wer auf seine Resultate nicht schwören kann, der mag immerhin zu seiner Übung analysiren, nur hüte er sich, seine Resultate als sicher zu veröffentlichen oder anzuwenden, es dürfte ihm nicht zum Vortheil, der Wissenschaft aber würde es nur zum Nachtheil gereichen."[8]

Words of the Master

(called „Hippocratic oath for analytical chemists"
by the historian of science William H. Brock)[8]

„Every analyst occasionally has doubts about the accuracy of his results, and also there are times when he knows his results to be incorrect. Sometimes a few drops of the solution were spilt, or some other slight mistake made. In those cases it requires strong conscience to repeat the analysis and not to make a rough estimate of the loss or apply a correction. Anyone not having sufficient will-power to do this is unsuited to analysis no matter how great his technical ability or knowledge. A chemist who would not take an oath guaranteeing the authenticity, as well as the accuracy of his work, should never publish his results, for if he were to do so then the result would be detrimental, not only to himself, but to the whole of science.[8]

2.3

Die Zeitschrift für Analytische Chemie

„... Nutzen, den sie auf die Entwickelung der Wissenschaft ausübt ...“[10]

Das Laboratorium mit Schülern, Auftragsanalysen, Methodenentwicklung, das Lehrbuchschreiben – man könnte denken, dass Remis Tage ausgefüllt waren. 1861 veröffentlicht er die Ankündigung einer Fachzeitschrift, die bis heute erscheint, und deren Redaktionsleitung bis 2001 über 4 Generationen hinweg bei seinen Nachkommen bis zum Urenkel lag. Ihr Herausgeber zu sein bedeutete fast tägliche Fron: Zuschriften von Autoren lesen, Wichtiges von Unwichtigem, Richtiges vom Falschen scheiden. Es heißt nicht nur, schlechte oder fehlerhafte Beiträge abzulehnen oder um Nachbesserung und Klärung von Unklarheiten zu ersuchen, in nicht wenigen Fällen auch Ergebnisse selbst experimentell zu überprüfen, sondern auch zu neuen interessanten Themen Autoren zu gewinnen.

The 'Journal of Analytical Chemistry'

'... the benefit that it brings to the development of science'[10]

With the laboratory and all the pupils, externally commissioned analyses, the development of his methods and the writing of textbooks, one might well imagine that Remi had no time left for any other work. On top of all that, however, in 1861 he announced that he would be bringing out a specialist journal. This journal is still published today, after having been edited by members of four Fresenius generations down to Remi's great-grandson. Editing the journal involved hard work for Remigius on an almost daily basis: reading submissions and distinguishing important articles from unimportant ones and right information from wrong. Nor did it only mean rejecting bad or faulty contributions and requesting

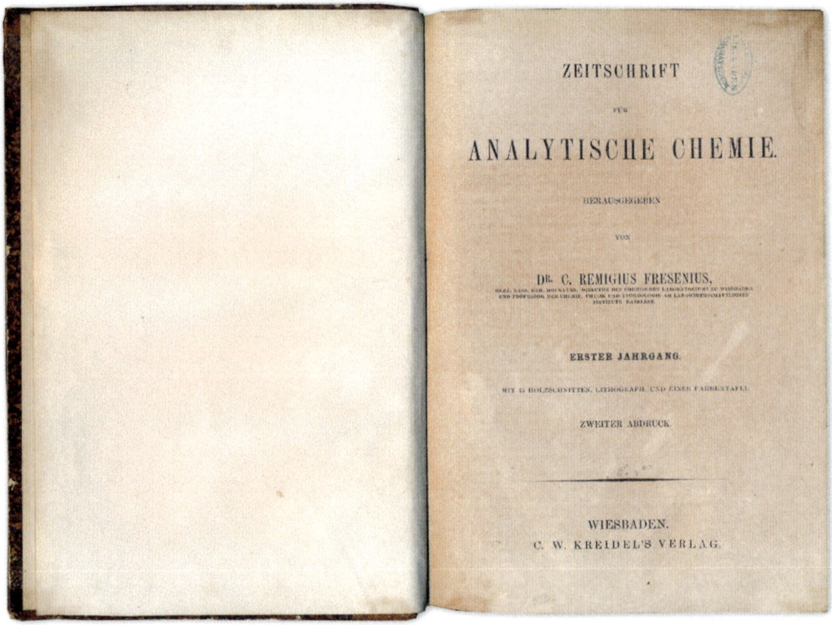

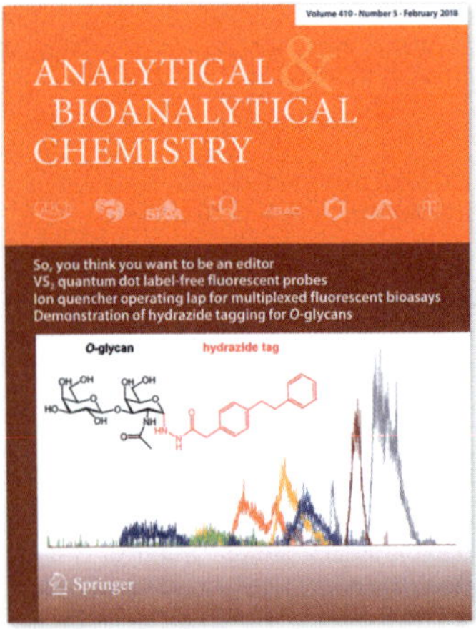

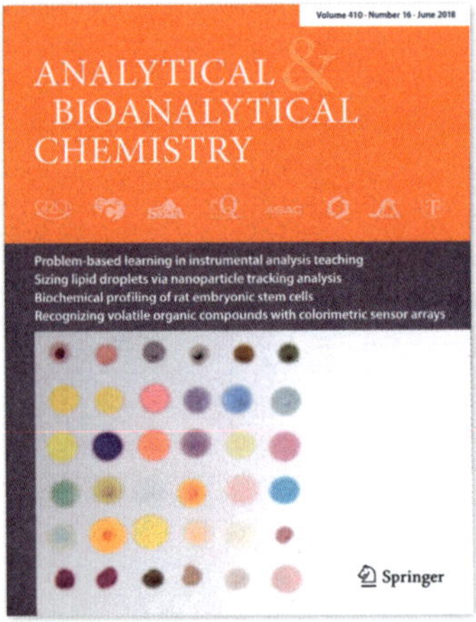

Die erste Ausgabe der von Fresenius gegründeten Zeitschrift für Analytische Chemie erschien (gedruckt bei Kreidel in Wiesbaden) 1862 (oben). In ununterbrochener Bandzählung heißt sie heute Analytical and Bioanalytical Chemistry und erscheint im Springer Verlag (unten).

The first volume of the analytical journal founded by Fresenius was published (printed at Kreidel's in Wiesbaden) in 1862 (above). In an uninterrupted series it is now called Analytical and Bioanalytical Chemistry published by Springer (below).

Warum lud sich Fresenius diese zusätzliche Aufgabe auf die offenbar breiten Schultern? Lesen wir ihn selbst:

> „Diese Pflege des Schatzes analytischer Methoden war bisher fast ausschliesslich den Lehrbüchern anvertraut. … aber in der Gegenwart, in welcher in fast allen Theilen der cultivirten Erde an dem Fortbau der Chemie rüstig gearbeitet wird, können sie allein die Aufgabe nicht mehr bewältigen … Diese Aufgabe kann nur eine periodische Schrift erfüllen … welche sich derselben ausschliesslich und ganz widmet …“[10]

Seine Überlegungen gelten heute umso mehr. Ab 1990 erschienen die Arbeiten der von da an „Fresenius' Journal of Analytical Chemistry“ genannten Zeitschrift nur noch in englischer Sprache. Seit 2002 führt der Springer Verlag die nun mit anderen europäischen analytischen Fachzeitschriften fusionierte Tradition unter dem Namen „Analytical and Bioanalytical Chemistry“ (ABC) fort. Ein nachhaltiger Ast am verzweigten Baum des „Alten“ bringt weiter frisches Grün. ●

authors to correct or clarify unclear points: in no small number of cases it also meant checking results through experiments on his own part, in addition to which Remi also had to cast around for authors to write on new and interesting subjects. Why did Fresenius load down his admittedly exceptionally broad shoulders with this additional task? Let us read his own words:

> 'The task of looking after the treasury of analytical methods has hitherto been almost exclusively performed in textbooks ... but in the present day, when progress in the field of chemistry is being industriously pursued in almost all parts of the civilized world, the books cannot fulfil this task on their own. ... This function can only be fulfilled by a periodical ... entirely and exclusively devoted to the task.'[10]

These thoughts are all the more valid today. In 1990 the periodical was renamed Fresenius' Journal of Analytical Chemistry and thenceforth published all its articles in English. Since 2002, the journal has lived on in the periodical Analytical and Bioanalytical Chemistry (ABC), which is a fusion of a number of European specialist analytical journals. The journal is a persistently healthy branch on the tree of the 'old master' that continues to bear fruit. ●

Zementproben der Firma Dyckerhoff. Aus der Sammlung Museum Wiesbaden

Concrete samples of Dyckerhoff company. Collection Museum Wiesbaden

Der Problemlöser

„... die materiellen Gesichtspunkte nicht außer Betracht zu lassen ...“[11]

The problem-solver

‘... not to leave material aspects out of consideration’[11]

3

Carl Remigius Fresenius hatte sich letztlich gegen die akademische Laufbahn entschieden und blieb zeitlebens der angewandten Forschung und der Anwendung ihrer Erkenntnisse verpflichtet.

Nach heutigen Begriffen würden wir Fresenius auch einen Dienstleister nennen. Er erhielt Aufträge staatlicher Behörden und aus der Privatwirtschaft. Sie beschränkten sich nicht auf reine Gehaltsanalysen – die häufig gutachterliche oder Schieds-Funktionen hatten und den Marktwert einer Ware (z.B. eines Erzes) oder ihren gesund-

After having finally decided against pursuing an academic career, Carl Remigius Fresenius remained for the whole of his life committed to applied research and to the application of his research findings.

In present-day terms we would also call Fresenius a service-provider. He received commissions from state departments and private enterprises. These were not restricted to commissions to analyse content, which were frequently used as part of expert judgments, or in

Mitarbeiter des Chemischen Laboratoriums Fresenius. Nach der Familientradition sollen von links nach rechts zu sehen sein: Carl Neubauer, Eugen Borgmann, Emil Erlenmeyer, Ernst Hintz. Atelierfoto.

Young staff of the Chemisches Laboratorium Fresenius. According to family tradition, from left to right: Carl Neubauer, Eugen Borgmann, Emil Erlenmeyer, Ernst Hintz.

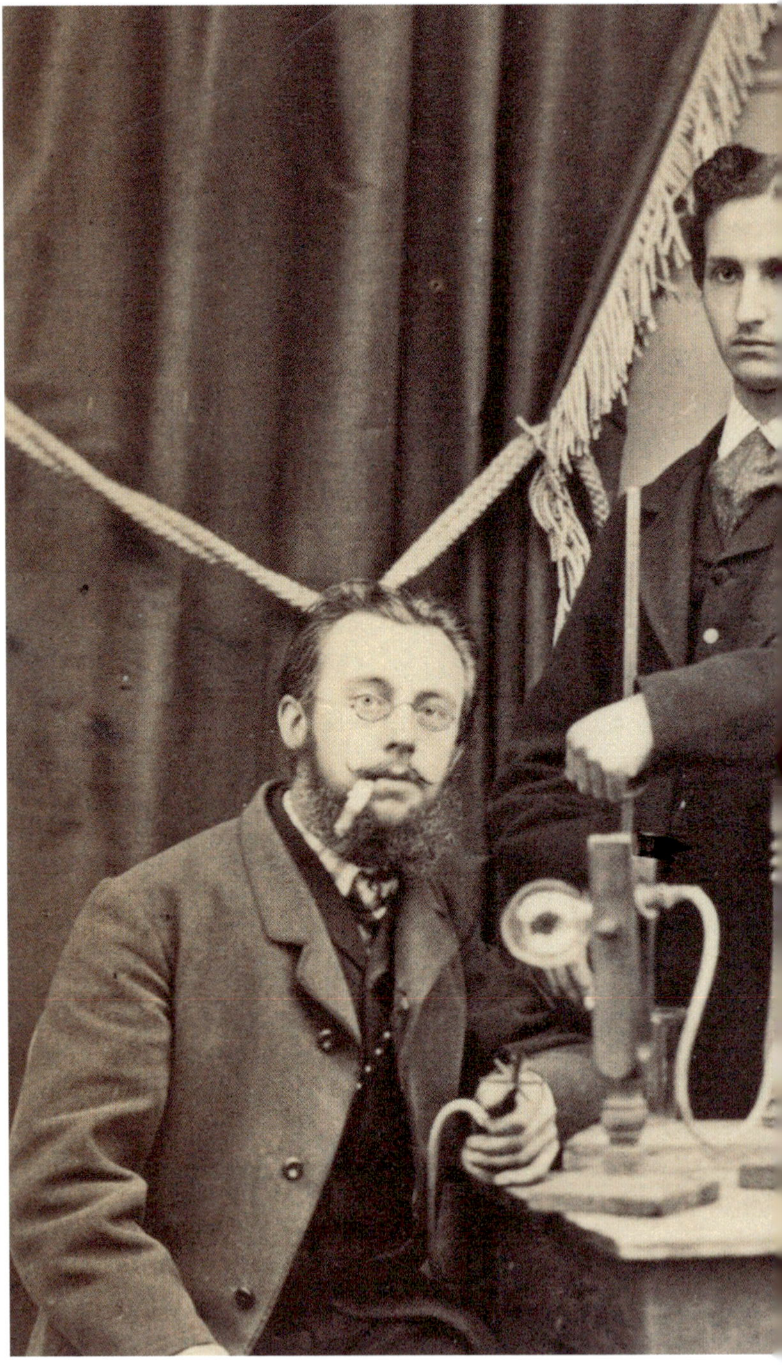

heitlichen Nutzen (z.B. eines Mineralwassers) feststellen halfen – sondern zielten vielfach darauf ab, den Gebrauchsnutzen eines Untersuchungsobjektes zu ermitteln. Wie witterungsbeständig sind Schieferplatten aus verschiedenen Vorkommen? Der Meister ließ sich etwas einfallen, um die Frage nicht erst nach jahrelanger Bewitterung beantworten zu können: Er hängte Probeplatten der Schiefer in eine unserem sauren Regen verwandte, jedoch höher konzentrierte Lösung von Sulfit (Schwefeldioxid) und beobachtete ihren Zerfall (oder ihre Haltbarkeit) über Wochen.

Die für analytische Arbeiten benötigten Reagenzien dürfen nicht verunreinigt sein, schon gar nicht mit dem Stoff, den man mit ihrer Hilfe bestimmen will. Fresenius beschreibt in seinen Lehrbüchern einfache Verfahren, ihre Reinheit zu prüfen. Der Merck-Chemiker Krauch beruft sich in der ersten Auflage seines Buches „Prüfung der chemischen Reagenzien auf Reinheit" von 1888 auf die Arbeiten von Fresenius. Der Hersteller Merck garantiert seitdem seinen Kunden nach standardisierten Methoden geprüfte Reinheit.

Der umfassenden Fachkenntnis des Analytikers Fresenius versicherten sich staatliche und wissenschaftliche Institutionen. Das wohl erste je erlassene Nahrungsmittelgesetz, das deutsche von 1879, half er als Mitglied zweier Kommissionen des Reichsgesundheitsamtes ebenso vorzubereiten wie Vereinbarungen der chemischen Fachgesellschaften zur Nahrungsmittel- und zur Weinanalytik. Aus der Erkenntnis, dass die gerade erst aus der Taufe gehobene Bakteriologie für Hygiene, Lebensmittel und Medizin von großer Bedeutung sei, schloss Fresenius auf die Notwendigkeit

the context of legal arbitration, or to determine the market value of a product (such as an ore) or its health benefits (as in the case of a mineral water); in the case of many commissions, the intention was simply to establish the practical utility of an object. What was the comparative weather-resistance, for example, of slate from various deposits? In order not to have to wait several years for an answer, Fresenius had the idea of hanging specimen slates in a sulphite solution (sulphur dioxide) similar to our acid rain but in a higher concentration and observed their decomposition (or durability) over a period of weeks.

Reagents employed in analytical work must not be contaminated, especially not with the substance that they are to be used to determine. In his textbooks, Fresenius describes simple methods for testing a reagent's purity. The Merck chemist Krauch draws upon the works of Fresenius in the first edition of his book on this very subject (Prüfung der chemischen Reagenzien auf Reinheit, 1888). Since then, the customers of the production company Merck have enjoyed a guarantee of purity based on standardized methods.

State bodies and institutions of scientific research made sure that they had Fresenius's expertise as an analyst at their disposition. As a member of two committees, Fresenius helped to prepare what was probably the first ever food law, Germany's Foods and Commodities Law of 1879, and furthermore played a part in the agreements of the expert chemistry societies on the analysis of food and wine. After having seen that the newly systematischer Kontrolle von Lebensmitteln, Umwelt und menschlichen Ausscheidungen – und eröffnete umgehend eine eigene Abteilung dafür.

Eine Gedenkschrift in den Berichten der Deutschen Chemischen Gesellschaft von 1919 fasst die Arbeit des Entwicklers analytischer und materialtechnischer Verfahren, qualitätssichernder Methoden, des problemlösenden Dienstleisters und des beratenden Experten so zusammen:

> „Das Laboratorium diente von Anfang an nicht nur dem Unterricht, sondern in nicht geringerem Umfange wissenschaftlicher und praktischer Arbeit. Es ist die Stätte, an der Fresenius und seine Mitarbeiter all die feinen Methoden ausgearbeitet haben, die heute Gemeinbesitz unserer Wissenschaft sind, an der ferner die von anderer Seite veröffentlichten analytischen Verfahren auf Grund experimenteller Nachprüfung kritisch gesichtet wurden, und an der schließlich das ganze Rüstzeug analytisch-chemischer Methodik in den Dienst praktischer Aufgaben gestellt wurde.“[12]

In diesem Dienst arbeiten mit modernen Methoden und an den Fragestellungen unserer Zeit SGS Institut Fresenius und Studierende und Ehemalige der Hochschule Fresenius in hunderten von Firmen weltweit. ●

born field of bacteriology was of great importance for hygiene, foods and medicine, Fresenius recognized the necessity of the systematic monitoring and control of foods, the environment and human excrement – and promptly opened a special department for this purpose.

A commemorative tribute in the 'Reports of the German Chemical Society' of 1919 gave the following summary of Fresenius's work in developing analytical procedures, technical procedures of a material kind and quality-control methods, and in acting as a problem-solving service-provider and expert advisor:

> 'From the very beginning, the laboratory served not only the purpose of teaching but also and to no lesser extent that of scientific and practical work. It is the place where Fresenius and his collaborators worked out all the refined methods that have now become common property of our science; there, furthermore, the analytical procedures published by other parties were examined through critical experimental testing and, finally, the whole methodological apparatus of analytical chemistry was applied in the service of practical tasks.'[12]

Today, the SGS Institut Fresenius, and students and alumni of the Fresenius University in hundreds of firms worldwide, are working in the same spirit of service, with modern methods, on important questions of our own time. ●

Das meinen Andere

„Gerade Anstalten, wie die hier fragliche, deren Begründer inzwischen nicht nur einen auf das Praktische gerichteten Sinn in fruchtbringender Weise bethätigt, sondern auch in der Zahl der Forscher eine hochgeachtete Stellung einnimmt, - haben wesentlich dazu beigetragen, die Wissenschaft der Chemie in das industrielle Leben ein- und überzuführen und damit der Wirtschaft unseres deutschen Volkes einen vorzüglichen Dienst geleistet.

Was speziell unseren Bezirk, in dem bekanntlich die chemische Industrie in höchst bedeutsamer und würdiger Weise vertreten ist, anlangt, so ist die überaus heilsame Einwirkung der Wiesbadener Anstalt unverkennbar.

An der Spitze einiger unserer namhaftesten Etablissements stehen Männer, welche als Jünglinge ihre naturwissenschaftliche Bildung in dem Laboratorium zu Wiesbaden empfingen.“[13]

What others think

‘Of all institutions, the one under consideration here – the founder of which not only uses his practically oriented mind in a highly fruitful manner, but also occupies a position of high esteem amongst researchers – has made a substantial contribution to introducing the science of chemistry into industrial life, thus doing a most excellent service to the economy of Germany.

In our district in particular, where the chemical industry is, as is well known, represented in a highly significant and worthy manner, the preeminently salutary effect of the Wiesbaden institution is unmistakable.

As young men, the directors of a number of our most renowned establishments received their formation as natural scientists in the laboratory at Wiesbaden.’[13]

3.1

Wasseranalytik

„Genaue Kenntnis der chemischen Beschaffenheit“[11]

Water analysis

'Precise knowledge of chemical composition'[11]

Am 20. April 1852 entsteigt in Schlangenbad im Rheingau-Taunus-Kreis ein hochgewachsener Mann von 34 Jahren mit langen, dunklen Haaren und gepflegtem ebensolchem Bart einer Kutsche. Vorsichtig lädt er eine Anzahl großer Glasflaschen und weitere Gerätschaften ab, darunter ein Thermometer. Er nimmt mehrere Quellen des Heilbades sorgfältig in Augenschein und notiert seine Beobachtungen. Dann taucht er nacheinander mehrere 6-Liter-Flaschen an der hintersten Quelle des unteren Kurhauses unter die Wasseroberfläche der Quelle ein und verschließt die randvollen Gefäße sorgsam mit gummiüberzogenen Korkstopfen. Was er damit bezweckt, hat er 1850 selbst beschrieben: Er ermittelt chemische Analysendaten nicht als „Zahlenfriedhof“, sondern um für

The date is 20 April 1852. A tall man 34 years of age with long dark hair and a well-trimmed beard alights from a coach at Schlangenbad in the Rheingau-Taunus district. Carefully, he unloads some large glass bottles and a number of instruments, including a themometer. He makes a close examination of several of the health resort's springs and makes notes of his observations. Then, one by one, he fills several six-litre bottles by submerging them in the waters of the spa's rearmost spring and then carefully seals them with rubber-covered cork stoppers. His goal in all this he had already described in 1850: the analytical data he is determining are not just lifeless statistics figures to be consigned to the bottom drawer but, rather, they are to be used to provide a die Einschätzung ihres wirtschaftlichen und gesundheitlichen Nutzens eine solide wissenschaftliche Basis zu liefern. Dazu entwickeln Fresenius und seine Nachfolger die Methodik laufend weiter.

Allein in Deutschland hat Carl Remigius Fresenius über 100 Quellen (davon etwa ein Drittel in Wiesbaden und der Taunusregion) zum Teil sehr umfassend untersucht[14], dazu zahlreiche weitere im Ausland. Bereits 1843 hatte er sowohl die warmen Quellen in Assmannshausen im Rheingau wie auch zwei Wässer von der Insel Java untersucht!

20 Jahre später gelang es Fresenius, die 28 Milliardstel Gramm Iodid, die in einem Kilogramm Selterswasser enthalten sind, zu bestimmen.[14] Dazu dampfte er 65 Kilogramm Wasser ein, bis nur noch die enthaltenen Salze übrig waren. Aus denen gewann er in mehreren Schritten einige Gramm eines Extrakts, das er auf Iodid analysierte. Sein Anteil im Wasser ist mit 28 mm Streckenlänge auf der Gesamtstrecke Köln-Wien vergleichbar. Fast nichts? Fresenius denkt weiter und meint, wenn er statt der 65 kg Wasser 100 oder gar 1000 kg eindampfen würde, müsste er auch Stoffe finden, die er wegen des geringen Gehalts noch nicht ermitteln kann.

Noch heute fallen uns zum Namen Fresenius leicht die Analysen ein, welche die Etiketten von Mineralwasserflaschen zieren und sich auf das SGS Institut Fresenius berufen. ●

solid scientific basis for the assessment of the waters' health benefits and economic potential. To this end, Fresenius and his successors constantly subjected their methods to further development.

Carl Remigius Fresenius investigated over a hundred springs In Germany alone (around a third of these being in Wiesbaden and the Taunus region), some very extensively[14], not to mention numerous others outside Germany. As early as 1843 he had investigated the warm springs of Assmannshausen, in the Rheingau, and two further waters from the island of Java.

Twenty years later Fresenius succeeded in quantifying the 28 parts of a billion of iodide contained in a kilogram of Selters mineral water.[14] In order to do this he boiled 65 kilograms of the mineral water dry until only the salts were left. From these, in several steps, he extracted a few grams of a substance which he analysed for iodide. Its proportion in the water can be compared to 28 mm as a fraction of the distance between Cologne and Vienna. Almost nothing? Practically negligible? Fresenius is more far-reaching in his thinking: if he was to boil down not just 65 but 100 or even 1000 kilograms of water, then he would certainly find substances that had previously not been identified on account of their minute quantities.

Today the name Fresenius is familiar to many from the analyses which derive from the SGS Institut Fresenius and adorn the labels of mineral water bottles. ●

Diese Glasflaschen aus der Zeit um 1880 enthielten Reagenzien, wie sie zur Mineralwasseranalyse verwendet wurden. Sammlung Friedrich

These glass vessels from the time around 1880 once contained reagents used for mineral water analysis. Collection Friedrich.

Hätte er sich das träumen lassen?

Was Fresenius im Fall des Iodids leistete, nennt man Spurenanalytik. Moderne Verfahren finden – wie er es voraussah – noch Mengen, die einem Hunderttausendstel des Iodidgehalts von Selterswasser entsprechen. Dafür muss man nicht tausende von Litern Wasser eindampfen – unsere modernen Verfahren sind viel empfindlicher als seine. Sie sind auch schneller und kostengünstiger: Für die Untersuchung einer Wasserprobe auf die Chlorid, Bromid, Nitrat, Nitrit und Sulfat brauchte das Laborteam von Remigius Fresenius einige Tage. Mit einem Ionenchromatographen lassen sich heute in 24 Stunden 80 Wässer analysieren.

Would he ever have dreamt of it?

What Fresenius achieved in the case of iodide was an instance of what we now know as trace analysis. Modern procedures are able – as he foresaw – to identify quantities as small as a hundred thousandth of the iodide content of Selters water. And as the modern methods are much more sensitive than his, they do not require the boiling down of thousands of litres of water. What is more, they are quicker and cheaper: while Fresenius's laboratory team took a number of days to test a sample of water for chloride, bromide, nitrate, nitrite and sulphate, with an ion chromatograph eighty different waters can be analysed in just one day.

3.2

Forensische Analytik

„Stützen für den Richterspruch“[15]

„Herz und Leber verbreiteten einen fauligen, widerlichen Geruch. Beide Organe waren ganz weich und liessen sich sehr leicht zerschneiden; ihre Farbe war dunkelbraun.“[16].

Das faulende Herz hatte bis zu seinem Ende das Blut von Johann Wirth aus Eschbach bei Usingen gepumpt. Wie so manch anderer Mensch hatte er sich öfters unwohl gefühlt, bevor er im Januar 1860 starb und begraben wurde. Hätte nicht sein Schwiegersohn vier Jahre später den Staatsanwalt mit einem Mordverdacht gegen Johanns Frau auf den Plan gerufen, hätte die Leiche weiter zu Erde werden können. Nun aber lagen die Organe auf dem Labortisch von Remigius Fresenius und er konnte darin „Dreifach-Schwefelarsen“

Forensic analysis

'Support for the judge's ruling'[15]

'Heart and liver emitted a putrid and nauseating smell. Both organs were quite soft and could be dissected very easily; they were dark brown in colour.'[16]

Until January 1860 the now rotting heart had pumped blood around the body of Johann Wirth from Eschbach near Usingen. Like many other people, he had often felt unwell before he died and was buried. If his son-in-law had not activated the state prosecutor four years later by raising a suspicion of murder against Johann's wife, the corpse could have continued its slow process of decomposition. Now, however, the organs lay upon the laboratory table of Remigius Fresenius, who was able to identify in them the presence of arsenic

(As_2S_3) nachweisen Nach weiteren Untersuchungen ist sich der vereidigte Gerichtsgutachter Fresenius sicher:
Es war ein Giftmord. Mehrfach dem Kaffee beigemischtes Arsen bewirkte nur Unwohlsein. Eine größere Menge in einer dicken Suppe, so die geständige Witwe, hatte Johann den Rest gegeben.[16]

Was befähigte Fresenius zum Gerichtsgutachter? Schon 1842 hatte er mit seinem Freund Lambert von Babo ein eigenes, „unter allen Umständen sicheres" Verfahren zur Arsenanalytik beschrieben.[17] Zu Recht maß er einer akribischen, peniblen, verlässlichen Methodik in der forensischen Chemie „hohe Bedeutung für das sociale Leben"[15] bei. So konnte er selbstbewusst schreiben, was jeder Gerichtsgutachter bis heute beherzigen wird: Er habe sich verpflichtet gefühlt „für Recht, Wahrheit und Wissenschaft nach Kräften einzustehen."[18] Für die moderne Analytik im Dienst der Verbrechensaufklärung gilt das noch heute. Ihre Fragestellungen sind jedoch andere: In Kriminalämtern untersuchen heutige Fresenius-Studierende Designer-Drogen auf winzige Anteile von Stoffen, die Hinweise auf die Ausgangssubstanzen und die mögliche Herkunft der hochgefährlichen, im Internet vielfach gehandelten Rauschmittel geben können. ●

trisulfide (As_2S_3). After further investigations the court-appointed expert speaking under oath is sure: it was murder by poison. On several occasions, a little arsenic in his coffee had only made him feel unwell. But, the widow confessed, a larger quantity mixed into a thick soup had done the trick admirably.[16]

What was it that qualified Fresenius to be appointed a court expert? Even back in 1842, together with his fiend Lambert von Babo, he had described a procedure for the analysis of arsenic that was 'certain in all circumstances'.[17] Quite rightly, he considered that a reliable and painstaking methodology in forensic chemistry was a matter of 'high importance for society'.[15] As a result, he was able to make a confident claim which will strike a chord with every court-appointed expert in our own day, namely, that he felt duty-bound 'to stand up with all his strength for justice, truth and science.'[18] This is equally valid for the modern analyses that are carried out with the goal of establishing responsibility for crimes. Now, however, the problems are different: in today's crime labs, students at Fresenius University examine designer drugs for tiny quantities of matter that may provide evidence of the original substances and of the possible origins of highly dangerous drugs very frequently bought and sold over the Internet. ●

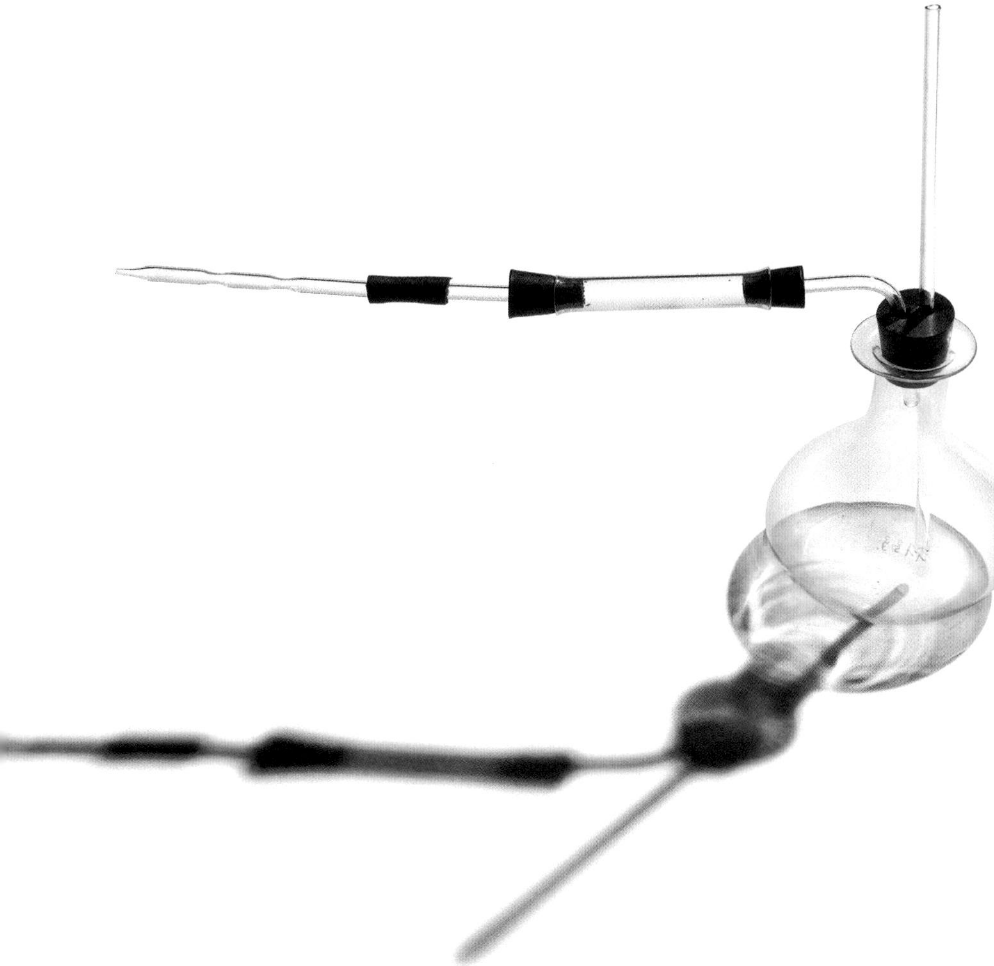

Chemische Apparatur zur Bestimmung des Arsens z.B. in Gerichtsfällen. Mit originalen und nachgebauten Bauteilen aus der Sammlung Friedrich, aufgebaut von Rüdiger Wandke.

Set-up for the quantitative determination of arsenic in, e.g., forensic cases. Made with original and reconstructed parts from the Friedrich collection, assembled by Rüdiger Wandke.

Merk-würdig

In einem Aufsehen erregenden Mordfall in Korneuburg bei Wien wies Fresenius 1859 gutachterlich nach, dass die von Kollegen als Mordbeweis angesehenen Untersuchungsergebnisse nicht auf eine Arsenvergiftung schließen ließen. Sie hatten Arsen mit dem ähnlich, aber anders reagierenden Antimon verwechselt, das damals in Arzneimitteln enthalten war. Die Hochschul- und Landesbibliothek in Wiesbaden bewahrt eine 250-seitige Dokumentation des Falls auf.[18]

Note-worthy

Following a murder in Korneuburg, near Vienna, that caused a widespread sensation in 1859, Fresenius was able to show through his expert report that the findings regarded by his colleagues as constituting evidence of murder did not in fact provide conclusive evidence of arsenic poisoning. His colleagues had confused arsenic with a similar substance that reacts differently and was at that time commonly contained in medication, namely, antimony. The case is recorded in a 250-page documentation preserved at the Hochschul-und Landesbibliothek in Wiesbaden.[18]

3.3

Lebensmittel- und Weinanalytik

„... ob der Wein rein oder verfälscht ist ...“[19]

Food and wine analysis

„... whether the wine is pure or adulterated ...“[19]

Dass Remigius Fresenius viele Wasseranalysen machte, wundert im quellenreichen Wiesbaden und Nassau nicht. Wiesbaden ist aber auch Teil einer Weinbauregion. Der Fresenius zugetane Herzog von Nassau hatte die Weinberge der Zisterzienser von Kloster Eberbach „geerbt“. Für seine staatliche Domäne machten Fresenius und sein Labor nicht nur Weinanalysen, sondern auch Studien zu Gärung, Weinbereitung und Düngung von Weinbergen. Dabei waren ihm der leider früh verstorbene Carl Neubauer und Eugen Borgmann unentbehrliche Mitarbeiter.[20]

Schon 1842 hatte ihn sein Lehrer Liebig zur Untersuchung der Salzbildung von Traubensäure angeregt. 1846 erhält er von der Herzoglichen Domäne „vorzügliche Weine“ zur Untersuchung und analysiert in

In an area so rich in springs as Wiesbaden and Nassau it is no surprise to find that Remigius Fresenius conducted so many water analyses. But Wiesbaden is also part of a wine-producing region. The Duke of Nassau, a great supporter of Fresenius, had ‘inherited’ the vineyards of the Cistercians of the abbey of Eberbach. For the Duke’s State Domains, Fresenius and his laboratory carried out not only wine analyses but also studies on fermentation, winemaking and the manuring of vineyards. In these endeavours he had indispensable collaborators in Carl Neubauer, who sadly died prematurely, and Eugen Borgmann.[20]

As early 1842 Fresenius’s teacher Liebig had encouraged him to investigate the salt formation of racemic acid. In

Zinkographien von Kaspar Kögler mit zugehörigem Text in einem Buch[21] über Malereien im Wiesbadener Ratskeller. Sie wurden im 2. Weltkrieg zerstört.

Zinkographies by Kaspar Kögler with accompanying text in a book[21] about paintings in the Wiesbaden Town Hall Cellar, destroyed in the Second World War.

Und des Weines Kennerſchaft
Stets verlör' an edler Kraft,
So daß manchesmal die Kenner
Stritten, ob der Stoff vom „Brenner",
Oder, ach du Kinderſpott!
Stamme von dem lieben Gott,
Drum als Leucht' in dieſer Wildnis
Malt' ich hier das hehre Bildnis
Unſeres braven alten Stahl,
Der jetzt kneipt im Götterſaal.
Kam ein Glas von ſ e i n e m Mund,
War's erforſcht bis auf den Grund;
Und das Faß, das er berochen,
Deſſen Urteil war geſprochen —

Denn vor seinem Urteilsschluß
Kroch ins Loch der Chemikus.
Drum ein Mann von solchem Holz
Sprach mit ganz gerechtem Stolz:

„Preist euren Stoff aus voller Lunge:
Der Kenner folgt der eignen Zunge!"

Ach es wäre schrecklich nett,
Wenn man auch so eine hätt';
Dann könnt' man mit frohen Mienen
Und Genuß das Geld verdienen,
Das der Handelsmann jetzt muß
Zahlen dem Fresenius.

— — — — — — — — — — — — —

Die vier Zwickel in der Höh
Sind für das Historische:

To illuminate this twilight
Here I portrayed as a highlight
Good old Stahl who, I can tell,
Now will on Mount Olympus dwell.
On his taste-buds wines were bound
To be probed down to the ground.
When a barrel he controlled,
Then its quality was told.
'Gainst the verdict of this hero
Chemist's odds went down to zero.

Thus, a man of such tough stock
Proudly tasted his good hock:
'Praise the wine with all your lung –
Connoisseurs follow their own tongue.'
I am sure no one would gripe,
Was his tongue of sim'lar type:
Then he would with mighty grin,
Savouring wine, the money win,
That without such well-trained senses
He'd owe Fresenius as expenses.

Merk-würdig

Der erste Dozent für Chemie und Physik der neu gegründeten Königlich-Preußischen Lehranstalt für Wein- und Obstbau war Fresenius' enger Mitarbeiter Carl Neubauer. Aus dieser Gründung ging die heutige Hochschule Geisenheim University hervor.

Zusammen mit der Hochschule Fresenius bietet diese in der internationalen Weinwelt renommierte Hochschule seit 2015 in einer bisher einzigartigen Partnerschaft einer staatlichen mit einer privaten Hochschule den Studiengang Lebensmittelsicherheit an.

Note-worthy

Fresenius's close colleague Carl Neubauer was the first teacher of chemistry and physics at the newly founded Royal Prussian School of Fruit- and Wine-Growing, out of which developed a present-day institution that enjoys great renown in the international wine world, namely, the Hochschule Geisenheim University.

Since 2015 - in a unique and unprecedented case of a partnership between a state and a private university - Geisenheim collaborates with the Hochschule Fresenius on offering the course of study 'Food Safety'.

der Folge auch Moste und verschiedene heimische Obstsorten. Sein Mitarbeiter Carl Neubauer unterwirft Weinlaub genauso einer Analyse wie nassverbesserte („gallisierte") Weine oder Weinbeeren in verschiedenen Reifestadien. Über mehrere Jahre hinweg entsteht unter der Leitung des Wiesbadener Labors eine deutsche Weinstatistik.

Der Mensch lebt jedoch nicht vom Wein allein – er kann auch Kartoffelschnaps trinken. Keine Fragestellung ist dem Meister zu unbedeutend! Dem Kartoffelbauern hilft ein einfaches, von Fresenius zusammen mit einem Herrn Schulze entwickeltes Verfahren zur Messung des Stärkegehalts und damit die Eignung zum Schnapsbrennen.

Auf die Gründung einer agricultur-chemisch-önologischen Abteilung 1868 folgte knapp zehn Jahre später der Beginn der wohl ersten deutschen systematischen Ausbildung von Lebensmittelchemikern. Ihre Schüler lernten erstmals die Mikroskopie auf Lebens- und Genussmittel anzuwenden. 1923 berichten die Söhne des Gründers, dass sie „regelmäßige Kontrolltätigkeit" für Kaffee Hag und Oetkers Backpulver übernommen haben.

Wer heute isst und trinkt, hat gute Chancen, vom Labor SGS Institut Fresenius untersuchte Lebensmittel zu genießen. Seine Fachleute stellen zum Beispiel sicher, dass keine mit krebserzeugenden Pilzgiften (Aflatoxinen) belasteten Nüsse in Süßigkeiten gelangen oder prüfen Lebensmittelverpackungen auf ihre Unbedenklichkeit. ●

1846 he analysed 'excellent wines' sent to him for this purpose by the ducal wine estates and also ciders and various local fruits. His colleague Carl Neubauer analysed not only vine leaves but also wines to which a sugar solution had been added during fermentation, and grapes at various stages of ripeness. Over a period of several years, a German wine statistics was developed under the direction of the Wiesbaden laboratory.

However, man cannot live from wine alone – he can, also drink potato schnapps! No problem was too banal for the Master's consideration. A procedure developed by Fresenius together with a man by the name of Schulze helped the potato-farmers by enabling them to measure their spuds' starch content and thus establish their suitability for distilling schnapps.

The foundation in 1868 of a special department for agricultural and oenological chemistry was followed ten years later by the launch of what probably was Germany's first systematic training course for food chemists. Its students were the first to learn to apply microscopy to foodstuffs, alcoholic drinks and tobacco. In 1923 the founder's sons reported that they had undertaken to conduct 'regular testing activities' for Hag coffee and for Oetker's baking powder.

Whoever eats and drinks today has a good chance of consuming foodstuffs analysed at the laboratories of the SGS Institut Fresenius, where experts for example ensure that no nuts bearing carcinogenic fungal toxins (aflatoxins) contaminate sweets and test food packaging to establish that it is innocuous. ●

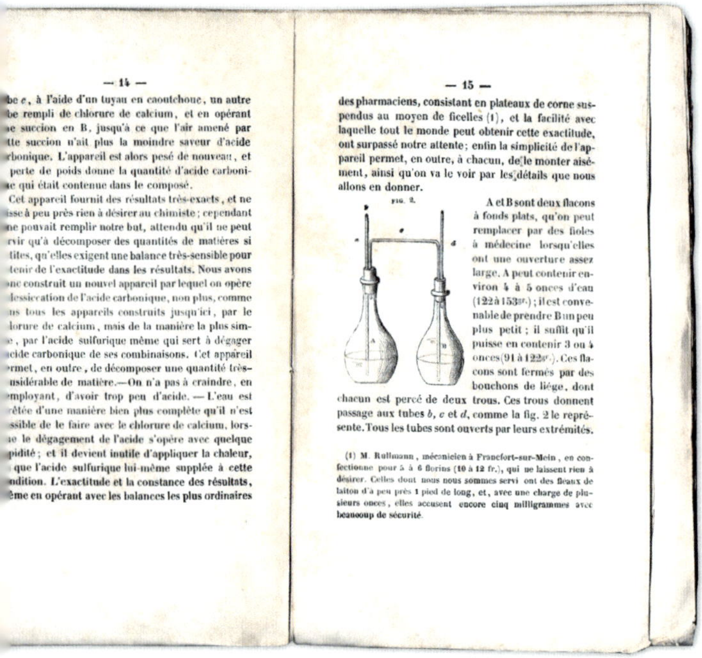
— 14 —

be *c*, à l'aide d'un tuyau en caoutchouc, un autre
be rempli de chlorure de calcium, et en opérant
ne succion en B, jusqu'à ce que l'air amené par
tte succion n'ait plus la moindre saveur d'acide
rbonique. L'appareil est alors pesé de nouveau, et
perte de poids donne la quantité d'acide carboni-
ue qui était contenue dans le composé.
Cet appareil fournit des résultats très-exacts, et ne
isse à peu près rien à désirer au chimiste; cependant
ne pouvait remplir notre but, attendu qu'il ne peut
rvir qu'à décomposer des quantités de matières si
tites, qu'elles exigent une balance très-sensible pour
tenir de l'exactitude dans les résultats. Nous avons
nc construit un nouvel appareil par lequel on opère
dessiccation de l'acide carbonique, non plus, comme
ns tous les appareils construits jusqu'ici, par le
lorure de calcium, mais de la manière la plus sim-
e, par l'acide sulfurique même qui sert à dégager
cide carbonique de ses combinaisons. Cet appareil
rmet, en outre, de décomposer une quantité très-
nsidérable de matière.—On n'a pas à craindre, en
mployant, d'avoir trop peu d'acide. — L'eau est
rêtée d'une manière bien plus complète qu'il n'est
ssible de le faire avec le chlorure de calcium, lors-
ue le dégagement de l'acide s'opère avec quelque
pidité; et il devient inutile d'appliquer la chaleur,
que l'acide sulfurique lui-même supplée à cette
ndition. L'exactitude et la constance des résultats,
ême en opérant avec les balances les plus ordinaires

— 15 —

des pharmaciens, consistant en plateaux de corne suspendus au moyen de ficelles (1), et la facilité avec laquelle tout le monde peut obtenir cette exactitude, ont surpassé notre attente; enfin la simplicité de l'appareil permet, en outre, à chacun, de le monter aisément, ainsi qu'on va le voir par les détails que nous allons en donner.

FIG. 2.

A et B sont deux flacons à fonds plats, qu'on peut remplacer par des fioles à médecine lorsqu'elles ont une ouverture assez large. A peut contenir environ 4 à 5 onces d'eau (122 à 153gr.); il est convenable de prendre B un peu plus petit; il suffit qu'il puisse en contenir 3 ou 4 onces (91 à 122gr.). Ces flacons sont fermés par des bouchons de liége, dont chacun est percé de deux trous. Ces trous donnent passage aux tubes *b*, *c* et *d*, comme la fig. 2 le représente. Tous les tubes sont ouverts par leurs extrémités.

(1) M. Rullmann, mécanicien à Francfort-sur-Mein, en confectionne pour 5 à 6 florins (10 à 12 fr.), qui ne laissent rien à désirer. Celles dont nous nous sommes servi ont des fleaux de laiton d'à peu près 1 pied de long, et, avec une charge de plusieurs onces, elles accusent encore cinq milligrammes avec beaucoup de sécurité.

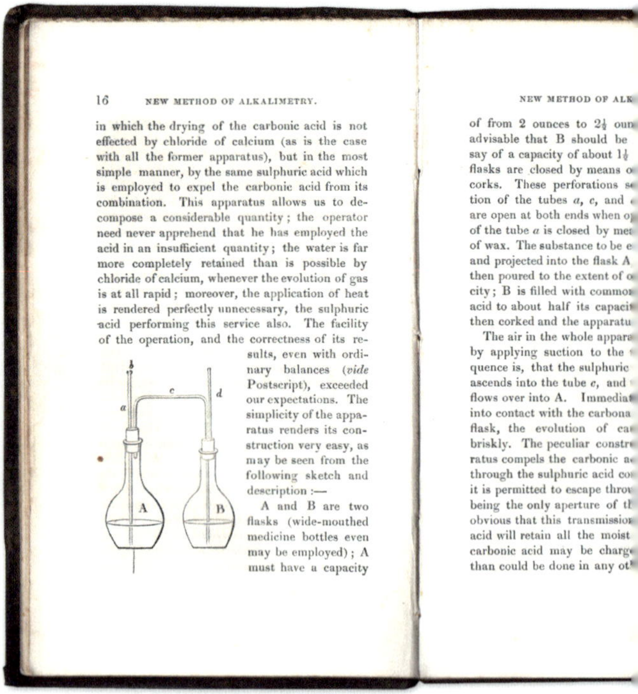
16 NEW METHOD OF ALKALIMETRY.

in which the drying of the carbonic acid is not effected by chloride of calcium (as is the case with all the former apparatus), but in the most simple manner, by the same sulphuric acid which is employed to expel the carbonic acid from its combination. This apparatus allows us to decompose a considerable quantity; the operator need never apprehend that he has employed the acid in an insufficient quantity; the water is far more completely retained than is possible by chloride of calcium, whenever the evolution of gas is at all rapid; moreover, the application of heat is rendered perfectly unnecessary, the sulphuric acid performing this service also. The facility of the operation, and the correctness of its results, even with ordinary balances (*vide* Postscript), exceeded our expectations. The simplicity of the apparatus renders its construction very easy, as may be seen from the following sketch and description :—

A and B are two flasks (wide-mouthed medicine bottles even may be employed); A must have a capacity

NEW METHOD OF ALK

of from 2 ounces to 2½ oun
advisable that B should be
say of a capacity of about 1½
flasks are closed by means o
corks. These perforations s
tion of the tubes *a*, *c*, and
are open at both ends when o
of the tube *a* is closed by me
of wax. The substance to be e
and projected into the flask A
then poured to the extent of o
city; B is filled with commo
acid to about half its capaci
then corked and the apparatu
The air in the whole appara
by applying suction to the
quence is, that the sulphuric
ascends into the tube *c*, and
flows over into A. Immedia
into contact with the carbona
flask, the evolution of ca
briskly. The peculiar constr
ratus compels the carbonic a
through the sulphuric acid co
it is permitted to escape thro
being the only aperture of t
obvious that this transmissio
acid will retain all the moist
carbonic acid may be charg
than could be done in any ot

Blick in die französische und englische Ausgabe eines Buchs von C.R. Fresenius und Heinrich Will über die Analyse von Industriechemikalien.

View of the French and English edition of a book written by C.R. Fresenius and Heinrich Will about the analysis of industrial chemicals.

3.4

Analyse technischer Produkte

„… daß man den richtigen Durchschnitt der Waare der Prüfung unterzieht …“[22]

The analysis of technical products

‘… that one should subject the right cross-section of the product to testing’[22]

Nein, so etwas kann sich niemand ausdenken: „Ueber die Ausfüllung des Mantels der feuerfesten Kassenschränke“ hat Carl Remigius Fresenius 1847 in den „Mittheilungen für den Gewerbeverein des Herzogthums Nassau“[23] berichtet. Es war nicht nur das Ziel des Meisters, „das ganze Rüstzeug analytisch-chemischer Methodik in den Dienst praktischer Aufgaben“[12] zu stellen – er tat es auch in einer bewundernswerten Breite und erweiterte die chemische Analytik um eine Vielzahl praxisnaher Testmethoden. Auf die Kassenschränke folgten Schwerspat und Anstreichfarben (1847), Tone (1848), Kalkstein (1851), Braunkohle und Braunstein (Manganerz) (1855) und schließlich Dachschiefer (1867). Arbeiten dieser Art schickte er an die herzogliche Verwaltung. Klappern gehört

No, nobody could make this up, even in their wildest imagination: in 1847, in the ‘Transactions of the Trade Association of the Duchy of Nassau’[23], Carl Remigius Fresenius published a report ‘On filling the cladding of fireproof safes’. For Fresenius, ‘to put the whole apparatus of chemical-analytical methodology in the service of practical tasks’[12] was not just a theoretical goal – he did it in practice, over an admirably broad range, and thus enriched chemical analysis with many very practical testing methods. The fireproof safes were followed by baryte and paints (1847), clays (1848), limestone (1851), lignite (brown coal) and pysolusite (manganese ore) (1855) and, finally, roofing slate (1867). He sent work of this kind to the ducal administration;

natürlich auch zum Handwerk eines Mannes, der sein Versprechen, seine Anstalt werde zur „Hebung“ der heimischen Wirtschaft beitragen, vielfältig einlöst.

In drei Publikationen in seiner eigenen Zeitschrift[24] entwickelt Fresenius spezielle Versuchsreihen, um – im Auftrag der Hersteller – Verfälschungen von Portland-Zement erkennen zu können. Dabei arbeitete er heraus, dass diese Aufgabe ohne eine Produktspezifikation des Industrieverbandes nicht lösbar ist.

Die technischen Untersuchungen reichen bis in sein letztes Lebensjahrzehnt. Wie kann man Rohstoffe beurteilen, aus denen Glühstrümpfe herstellt werden? Schon ein Jahr nach ihrer Erfindung durch den österreichischen Chemiker Carl Auer von Welsbach untersuchte Fresenius mit seinem Schwiegersohn diese Frage und kam zum Schluss, dass nicht der Thorium- sondern der Cergehalt des Materials eine für die Leuchtkraft entscheidende Rolle spielt.[25] Übrigens stellte man in Wiesbaden 1892 die ersten 70, im Folgejahr dann weitere 380 Gaslaternen auf. Das Chemische Laboratorium Fresenius ist technisch ganz vorn mit dabei. So ist es bis heute: SGS Institut Fresenius beteiligt sich zum Beispiel an Warentests, untersucht Konstruktionsmaterialien auf Materialfehler und prüft elektronische Bauteile auf ihre Korrosionsfestigkeit. ●

self-advertisement was naturally all part and parcel of the business of a man who had given – and was intent upon keeping – his word that his institution would make a substantial contribution to raising the level of the local economy.

In three articles in his own journal[24], Fresenius presented a number of special series of experiments designed, on commission from its producers, to identify fake Portland cement. In so doing he established that the task cannot be fulfilled without a product specification issued by the relevant industrial association.

His technical investigations continued into the last decade of his life. How can one assess the raw materials from which incandescent mantles are produced? A year after the gas mantles had been invented by the Austrian chemist Carl Auer von Welsbach, Fresenius investigated this question with his son-in-law and came to the conclusion that it was not the thorium but the cerium content of the material that was the critical factor for the level of luminosity.[25] In passing, it may be noted that in 1892 Wiesbaden erected its first seventy gas lamps, which were followed in 1893 by a further 380. The Fresenius Chemical Laboratory played a major technical role in this context. This is still the case today, when the SGS Institut Fresenius, among its many other undertakings, is involved in product-testing, in testing construction materials for material defects, and in testing electronic components for their resistance to corrosion. ●

über den Nachweis von Verfälschungen im Portland-Cement. 177

Es sind dies:

a) das specifische Gewicht,
b) der Glühverlust,
c) das Verhalten zu Wasser, respective die Alkalinität der Wasserlösung,
d) das Verhalten zu verdünnter Säure,
e) das Verhalten zu Chamäleonlösung,
f) das Verhalten zu gasförmiger Kohlensäure.

a. Das specifische Gewicht wurde nach der Methode bestimmt, welche Herr Dr. Schumann zu Amöneburg im Jahre 1883 auf der Generalversammlung des Vereins deutscher Cement-Fabrikanten in Berlin mitgetheilt hat.*) Dieselbe beruht darauf, dass man den durch eine gewogene Quantität des Cementes eingenommenen Raum ermittelt, indem man den Cement in ein mit Terpentinöl gefülltes Gefäss bringt und beobachtet wieviel das Niveau der Flüssigkeit steigt.**) Der zu diesen Bestimmungen angewandte Apparat ist in Fig. 21 abgebildet, er besteht aus einem etwa 100 bis 150 *cc* fassenden Gefässe und einem mit Hülfe eines Glasschliffes eingesetzten Rohre, welches 40 *cc* fasst und in $^1/_{10}$ *cc* eingetheilt ist. Der Apparat wird beim Gebrauch bis zum Nullstrich der Theilung mit Terpentinöl gefüllt, dann bringt man den gewogenen Cement mit Hülfe eines weiten Trichters durch das eingetheilte Rohr nach und nach in den Apparat, wartet bis derselbe sich so weit abgesetzt hat, dass man den Flüssigkeitsstand ablesen kann und erfährt in der verdrängten Anzahl von Cubikcentimetern das Volumen des Cementes. Man hat dann nur mit dieser Zahl in das Gewicht zu dividiren um das specifische Gewicht zu erhalten.

Fig. 21.

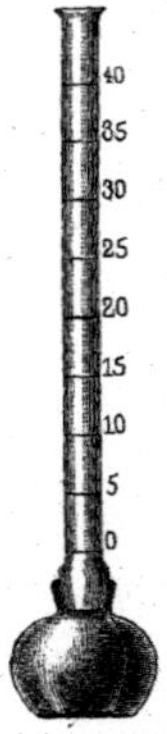

Die Methode erwies sich im Vergleich zur Bestimmung mittelst eines mit Terpentinöl gefüllten, nicht zu kleinen Pyknometers als viel

*) Protokoll der Verhandlungen des Vereins deutscher Cementfabrikanten am 22., 23 und 24. Februar 1883 p. 47. Berlin Druck von Julius Fenske.

) Das Princip ist dasselbe, welches auch schon von Phipson (diese Zeitschrift **1, 452), Brügelmann (diese Zeitschrift **21**, 178) und von Gisevius (diese Zeitschrift **23**, 51) bei speciellen Methoden der specifischen Gewichtsbestimmung benutzt worden ist.

12*

Ausschnitt aus der ersten Publikation über die Untersuchung der Verfälschungen von Portlandzement von 1885[24a].

Page from the first publication on the analysis of adulterated Portland concrete 1885[24a].

Worte des Meisters

„Die in Nachstehendem beschriebenen Versuche sind ausgeführt auf Veranlassung des Vereins deutscher Cementfabrikanten, der uns um Ausarbeitung von Methoden ersuchte, die es möglich machen einen durch Zumischung eines fremden Körpers verfälschten Cement als solchen zu erkennen, da der Verein bei seinem Auftreten gegen die immer mehr überhand nehmende Verfälschung nur dann auf einen Erfolg hoffen konnte, wenn Methoden zur Entdeckung der fremden Zusätze existirten.“[24a]

Words of the Master

‘The experiments described below are carried out at the instigation of the Association of German Cement-Producers, which requested us to develop methods for recognizing cements that have been adulterated through the admixture of a foreign body. The reason was that the Association can only hope for success in its fight against falsification, which is growing and increasingly gaining the upper hand, if methods exist through which the foreign admixtures can be identified.’[24a]

3.5

Physiologische Analytik und Mikrobiologie

„Bedeutung für die Untersuchung von Trinkwasser, Nahrungs- und Genußmitteln, Luft …“[26]

Physiological analysis and microbiology

‘Significance for the investigation of drinking water, foodstuffs and alcoholic drinks and tobacco, air …’[26]

Stark vereinfacht kann man sagen: Krankmachende Moleküle untersucht die klinische oder physiologische Analytik, zum Beispiel in Körpersäften wie Blut und Urin. Krankmachende Organismen, vor allem Bakterien, untersucht die Mikrobiologie. Beide Felder hatten früh einen Platz im Fresenius'schen Laboratorium.

Fresenius' kompetenter Mitarbeiter Carl Neubauer ging nicht nur dem Wein (» Kapitel 3.3) analytisch auf den Grund – er verfolgte auch seinen Weg bis zum Ende. Daraus wurde schließlich ein umfangreiches Standardwerk zur Harnanalyse.[27] Schon 1856 hatte er in Gärung übergehenden Harn von Zuckerkranken untersucht. Heute wird bei der Diabetesdiagnostik Mdie Tatsache genutzt, dass bei einem über Monate hohen Zuckerspiegel der rote

To put it in very simple terms, while molecules that cause illness are investigated by clinical or physiological analysis, notably body fluids such as blood and urine, organisms that cause illness, in particular bacteria, are investigated by microbiology. Both fields were represented at an early stage in Fresenius's laboratory.

Fresenius's able collaborator Carl Neubauer not only analysed wine (» chapter 3.3) but also followed its path through to the bitter end, finally publishing a substantial standard work on the analysis of urine.[27] As early as 1856 he had investigated diabetics' urine at the stage of incipient fermentation. Present-day diagnosis of diabetes makes use of the fact that when the blood sugar level remains high over a period of months, the haemo-

Blutfarbstoff Zuckermoleküle bindet. Ein Gerät der Firma Diasys misst diese HbA1c genannten Moleküle effizient. An seiner Entwicklung waren ehemalige Studierende der Hochschule Fresenius beteiligt.

„Chemie im Essen" macht krank? Lebensmittelchemiker sagen, dass Erkrankungen durch Mikroorganismen die weitaus häufigste Ursache für Schädigungen sind. In den siebziger Jahren des 19. Jahrhunderts erschienen wegweisende Veröffentlichungen zur Mikrobiologie, allen voran die von Robert Koch. Wenig später erkannte Fresenius die Bedeutung dieses neuen Arbeitsgebietes und stellte den Koch-Schüler Ferdinand Hüppe als Leiter einer von ihm eingerichteten hygienisch-bakteriologischen Abteilung ein. In den 5 Jahren seiner Tätigkeit in Wiesbaden erschien sein praxisnahes Buch „Methodik der Bakterienforschung"[28a]. Einzelphänomene wie die in Finthen und Gonsenheim aufgetretenen Cholera-Erkrankungen bearbeitete er ebenso wie den „Zusammenhang der Wasserversorgung mit der Entstehung von Infektionskrankheiten und die hieraus in hygienischer Beziehung abzuleitenden Folgerungen"[28b]. Die mikrobiologische Abteilung im heutigen SGS Institut Fresenius untersucht weiterhin Trink- und Mineralwasser, Lebensmittel und technische Produkte, prüft darüber hinaus Kosmetika, Arzneimittel und Bedarfsgegenstände und ist mit der Hygiene-Überwachung von Bädern, Hotels und Gaststätten betraut. Zur Identifizierung von Mikroorganismen verwendet sie auch Geräte und Diagnostica der von dem Fresenius-Schüler Marcel Mérieux aus Frankreich (» Kapitel 4) gegründeten Firma BioMérieux. •

globin binds sugar molecules. These molecules, known as HbA1c, can be measured efficiently with an instrument from the firm Diasys, which was developed with the involvement of a number of alumni of the Fresenius University of Applied Sciences.

Does 'chemistry in foods' cause illness? According to food chemists, the human body most frequently suffers harm from illnesses caused by microorganisms. The 1870s saw the publication of numerous trail-blazing works on microbiology, first and foremost those of Robert Koch. A short time later, Fresenius recognized the importance of this new field and appointed a doctoral student of Koch's, Ferdinand Hüppe, as director of a department of hygiene and bacteriology that he, Fresenius, had set up. Hüppe worked in Wiesbaden for five years and during this period published his practically oriented book 'The Methods of Bacterial Research'[28a]. Hüppe not only dealt with individual phenomena such as the outbreaks of cholera in Finthen and Gonsenheim but also considered such subjects as the 'connection between the water supply and the generation of infectious diseases, and the hygiene-related conclusions to be drawn from it'[28b]. Today, the department of microbiology at the SGS Institut Fresenius continues to investigate drinking and mineral water, foods and technical products; in addition it tests cosmetics, medicines and sanitary requisites and monitors hygiene at baths, hotels and restaurants. For the identification of microorganisms, SGS also uses apparatus and reagents of BioMérieux, a company once founded by the French Fresenius student Marcel Mérieux (» chapter 4). •

Mikroskop 54010 Zeiss AHJ 265, um 1880. Sammlung Friedrich

Microsope 54010 Zeiss AHJ 265, around 1880. Collection Friedrich

Das meinen Andere

„Allen Fortschritten der Wissenschaften folgte er [C.R. Fresenius] und schuf sachgemäße Einrichtungen, um sie in den Unterricht einbeziehen zu können. So errichtete er vor allem, noch bevor an irgend einer deutschen Universität Ähnliches vorhanden war, an seinem Laboratorium 1884 eine Stätte für bakteriologische Unterrichtskurse und Untersuchungen."[29]

In the opinion of others

'He [C.R. Fresenius] followed all the progress being made in the sciences and set up appropriate structures in order to be able to incorporate the progress in teaching. Above all, in 1884, before anything similar had been established at any German university, he created at his laboratory a place for bacteriological courses and investigations.'[29]

Der Lehrer

„Den Forschungstrieb zu entfachen …“[30]

The teacher

‘To kindle the drive to do research …’[30]

4

„Welch ein Lehrer war aber auch Fresenius! … Er erfüllte seine Schüler mit dem wahren Geiste wissenschaftlicher Forschung, mit dem Streben nach Wahrhaftigkeit. Niemals duldete er ein ödes mechanisches Abarbeiten einer festgelegten Vorschrift; das geistige Erfassen, die Erziehung zum selbständigen Denken war ihm Anfang und Ende des Unterrichts.“[31]

‘But what a teacher Fresenius was! … He filled his students with the true spirit of scientific research, with the will to strive for truthfulness. Never did he tolerate the mindless, mechanical carrying out of a predetermined procedure; intellectual understanding and training in independent thought, these were for him the alpha and omega of teaching.’[31]

Die sich ihm und seinen Nachfolgern als Schüler anvertrauten, waren zahlreich: In

A huge number of pupils were entrusted to his and his successors’ care: according to den ersten 50 Jahren (also bis kurz nach des Meisters Tod) verzeichnet die Statistik etwa 2500, bis 1948 dann insgesamt knapp 5000 Schülerinnen und Schüler. Von ihnen kamen zu seinen Lebzeiten 35 %, in den 50 Jahren nach seinem Tod (in einer Zeit mit zwei Weltkriegen) immer noch 23 % aus dem Ausland. Die Herkunftsangaben in den Listen umfassen fast alle europäischen Länder, darunter mit besonders hohem Anteil England und Schottland, Frankreich, Polen, Russland, Schweden, die Schweiz, Tschechien sowie in Übersee die USA mit weit über 100 Studierenden. China ist zu Beginn des 20. Jahrhunderts auch stark vertreten. Die Reichweite von Fresenius' Bildungstätigkeit lässt sich aber nicht nur in Quantitäten messen. Schon in seiner Zeit am Landwirtschaftlichen Institut hatte Fresenius dem Herzog von Nassau, seinem prominenten Förderer, Vorlesungen über Chemie gehalten. Richtig interessant wird es aber, wenn man die vollständig erfassten Namen der Absolventen und (seit 1908 auch) Absolventinnen nach bekannten und weniger bekannten Namen aus Wissenschaft und Wirtschaft durchsucht. Unter ihnen waren viele junge Leute aus der Gründergeneration der chemischen Industrie, oder deren Söhne und Enkel: Wilhelm, Ludwig und Carl Emmanuel Merck, Wilhelm Heraeus, Carl und Otto Leverkus, Eugen Lucius und Adolf Brüning (der Gründer und ein Teilhaber der Firma Hoechst) und Eugen de Haën. In der Rhein-Main Region bekannte Schüler des Laboratoriums waren unter anderen Wilhelm Kalle, Rudolf Koepp und Fritz Geromont. Der Mitentdecker von und weltweit erste Fabrikant von Saccharin, als

the statistics, in the first 50 years (that is to say, until shortly after the Master's death) the pupils amounted to around 2,500 and by 1948 had grown to a total of almost five thousand. In his lifetime 35 per cent of these came from abroad, and 23 per cent in the 50 years after his death (in spite of the period including two world wars). The indications of country of origin in the lists cover almost all the lands of Europe, with particularly high proportions coming from England and Scotland, France, Poland, Russia, Sweden, Switzerland and Bohemia (todays Czech Republic), not to mention overseas countries such as the United States with well over a hundred students. At the beginning of the twentieth century, China was also well represented. However, mere numbers cannot do full justice to the scope of influence of Fresenius's teaching activities. As early as in his time at the Agricultural Institute, Fresenius had delivered lectures on chemistry to his eminent patron the Duke of Nassau. Moreover, it really begins to get interesting when one combs the complete list of graduates (which from 1908 includes women) for famous and less well-known names from the worlds of science and commerce. Amongst these are found many young members of the chemical industry's founder generation, or indeed their sons and grandsons: Wilhelm, Ludwig and Carl Emmanuel Merck, Wilhelm Heraeus, Carl and Otto Leverkus, Eugen Lucius and Adolf Brüning (the founder of and a partner in the company Hoechst), and Eugen de Haën. The laboratory's students from the Rhine-Main region included Wilhelm

Porträt von Emil Erlenmeyer um 1863. Der in Wehen geborene Chemiker war 1848 erster Assistent von Carl Remigius Fresenius. Später wurde der Erfinder des nach ihm benannten Kolbens Professor in Heidelberg und München. Zu seinen Schülern zählen der Chemiker und Musiker Alexander Borodin und die Nobelpreisträger Buchner, Willstätter und Hans Fischer.

Portrait of Emil Erlenmeyer around 1863. Born in Wehen im Taunus, the chemist was Fresenius' first lab assistant in 1848. The inventor of the flask which bears his name later became a professor in Heidelberg and München. Among his students we find the chemist and composer Alexander Borodin and the Nobel prize winners Buchner, Willstätter and Hans Fischer.

Färberei der Firma Kalle. Der Gründer der Fima, Wilhelm Kalle (1838-1919) war 1857/58 Schüler und 1859/60 Assistent im Chemischen Laboratorium Fresenius. Er ist einer von zahlreichen Persönlichkeiten der chemischen Industrie des 19. Jahrhunderts, die ihren Berufsweg mit einer Ausbildung bei Fresenius begannen oder förderten.

Dye factory of Kalle. The companiy's founder, Wilhelm Kalle (1838-1919) had been a student 1857/58 and a lab assistant 1859/60 in Fresenius' laboratory. He is one of the many figures from the 19th century chemical industry who started or developed their career with a training under Fresenius.

Die Lehrbücher zur qualitativen (17 Auflagen) und zur quantitativen (7 Auflagen und zahlreiche Nachdrucke) Analyse erfuhren zahlreiche Übereetzungen. Ein Ausschnitt aus langen Regalreihen.

The textbooks on qualitative (17 editions) and quantitative (7 editions with many reprints) analysis ere frewuently translated. Section of a large shelf.

Süßstoff, Constantin Fahlberg, genoss ebenso eine gründliche Schulung beim international geachteten Meister wie Marcel Mérieux aus Lyon, Gründer der heute bioMérieux genannten Diganostikfirma und die Söhne des Gründers von Mallinckrodt in den USA. In der Wissenschaft ragen der erste Assistent von Fresenius, Emil Erlenmayer und der Nobelpreisträger Emil von Behring heraus. Übrigens: Wer denkt schon an Fresenius beim Genuss eines Cocktails mit Angostura-Bitter? Der nach Venezuela ausgewanderte Erfinder der ursprünglich für seine magenkranke Schwiegermutter entwickelten Spirituose schickte seinen Sohn nach Wiesbaden. Mehrere schwedische Fachleute für Hygiene und Lebensmittel studierten in Wiesbaden, bevor sie das Erlernte in ihrem Heimatland an vorderster Front fruchtbar umsetzten. Die Liste ließe sich lange fortsetzen, das erschlossene Material harrt noch der weiteren Auswertung.

Was machte die Anziehungskraft dieser Bildungsstätte aus? Erstens: Sie lehrte gründlich und zum eigenen Durchdringen des Stoffes anregend und hielt zum produktiven statt zum reproduktiven Lernen an. Zweitens: Sie lehrte praxis- und anwendungsnah, weil im gleichen Laboratorium oft die gleichen Leute lehrten, forschten und Auftragsanalytik machten. Sie lehrten nicht nur Wissen, sondern auch Können und Anwenden. Die Hochschule Fresenius hat 200 Jahre nach der Geburt des Gründers 900 Mitarbeiterinnen und Mitarbeiter und 12000 Studierende in den 5 Fachbereichen Chemie & Biologie, Gesundheit & Soziales, Wirtschaft & Medien, Design und Online-plus. Bis auf den

Kalle, Rudolf Koepp and Fritz Geromont. The co-discoverer and first manufacturer of saccharine (as a sweetener), Constantin Fahlberg, enjoyed just as thorough a schooling from the internationally renowned teacher Fresenius as did Marcel Mérieux of Lyon, founder of the diagnosis firm now operating under the name bioMérieux, or as the sons of the founder of Mallinckrodt Pharmaceuticals in the USA. Two outstanding figures in the scientific field were Fresenius's first assistant, Emil Erlenmayer, and the Nobel prizewinner Emil von Behring. By the way: who thinks of Fresenius when enjoying a cocktail containing Angostura bitters? But there is a connection. The inventor, a German who had emigrated to Venezuela and originally developed the bitters to alleviate his mother-in-law's stomach complaints, sent his son to Wiesbaden. The pioneering and highly fruitful work done by several food and hygiene experts in Sweden resulted from them having studied in Wiesbaden. Many more such cases could be listed; in addition, there is still an ample fund of documentation and evidence which is waiting to be further evaluated.

What was it that gave this educational establishment its particular power of attraction? Firstly, the teaching was thorough, it was aimed at stimulating students to be proactive in acquiring a thorough understanding of the subject matter, and its priority was productive rather than reproductive learning. Secondly, because the teachers often also did research and conducted externally commissioned analyses at their laboratories, they not only imparted knowledge

Das meinen Andere

“Im Frühjahr 1864 reisten Edward und Otto Mallinckrodt nach Deutschland um unter der Anleitung von Professor Karl Remigius Fresenius, des berühmten „Vaters der chemischen Analyse“ zu studieren ... machten anschließend Praktika ... und kehrten 1867 nach Missouri zurück. Sie hatten die beste Ausbildung erhalten, die die Welt zu bieten hatte. Vor allem aber hatten sie sich das akribische Qualitätsstreben ihrer Gastgeber zu Eigen gemacht ...“[32]

In the opinion of others

‘In the spring of 1864, Edward and Otto Mallinckrodt traveled to Germany to study … under the tutelage of Professor Karl Remigius Fresenius, renowned “father of chemical analysis” … [They] served practical apprenticeships … [and] returned to Missouri in 1867, having received the finest training the world had to offer. More importantly, they had also assimilated their German hosts’ painstaking devotion to quality.’[32]

heutigen Tag gilt für die Hochschule Fresenius, was Remigius Fresenius anlässlich der Gründung seiner pharmazeutischen Lehranstalt schrieb:

> „Die Anstalt wird ihren Zweck nur erfüllt glauben, wenn es ihr gelingt, die jungen Studierenden nicht allein mit gediegenen Kenntnissen auszurüsten, sondern in ihnen auch Lust und Liebe zur Wissenschaft und Forschungstrieb zu entfachen, weil nur darin die Mittel liegen, auch im späteren Leben mit den in rascher Entwicklung begriffenen Naturwissenschaften überhaupt fortschreiten zu können."[30] ●

but also taught skills and applications. Two hundred years after its founder's birth the Fresenius University of Applied Sciences has a staff of 900 and 12,000 students in its five faculties: Chemistry & Biology, Health & Social, Economics & Media, Design, and 'online-plus'. Even today, what Remigius Fresenius wrote at the foundation of his school of pharmaceutics is still put into practice at Fresenius University:

> 'The institution will only be able to consider it has achieved its purpose if it succeeds not only in equipping the young students with a solid fund of knowledge, but also in kindling in them a love of and enthusiasm for science and a drive to do research, because only these later enable them to make any progress in the rapidly developing field of the natural sciences.'[30] ●

Wo kamen die Fresenius-Studenten 1848–1948 her?

Where did Fresenius students 1848–1948 come from?

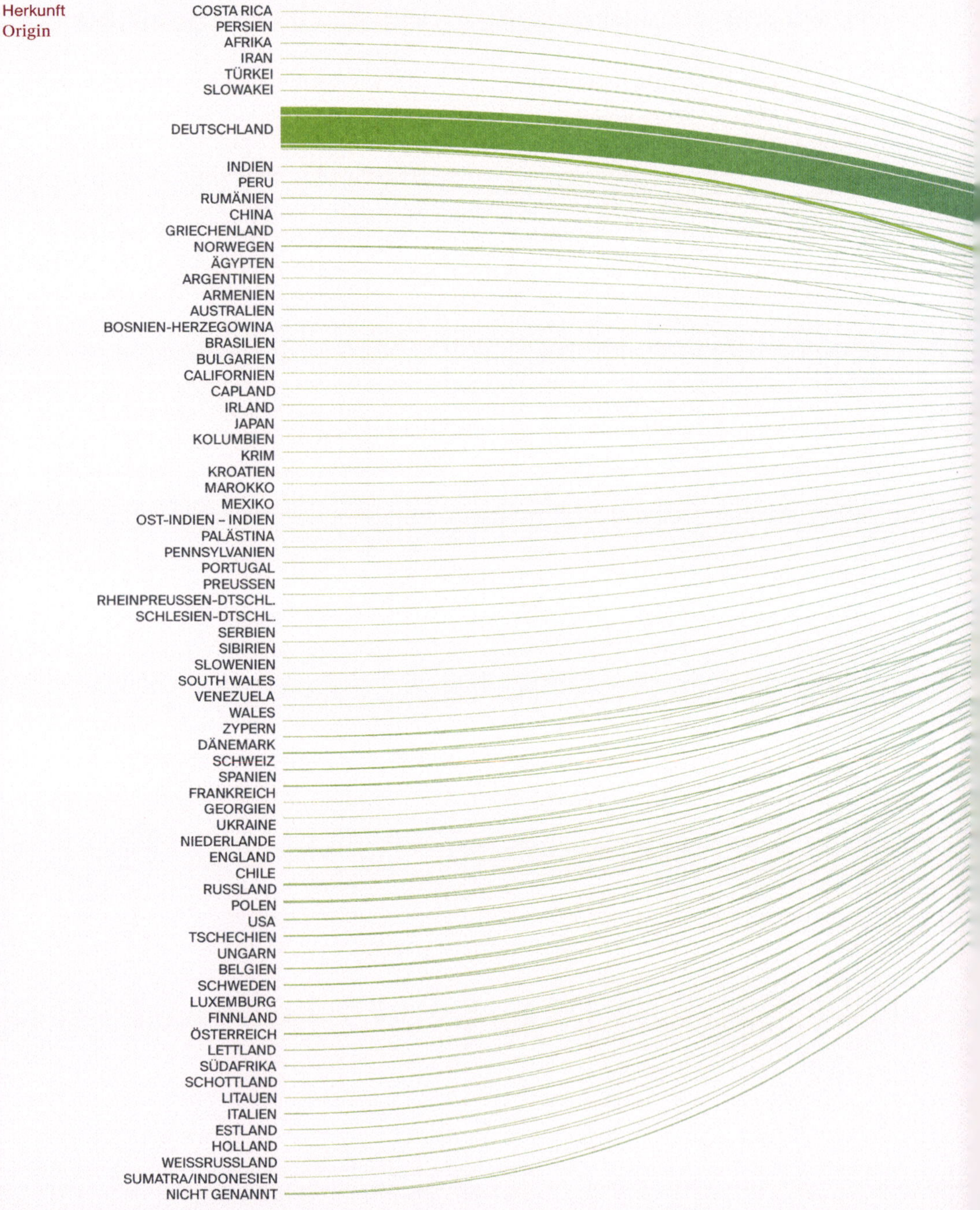

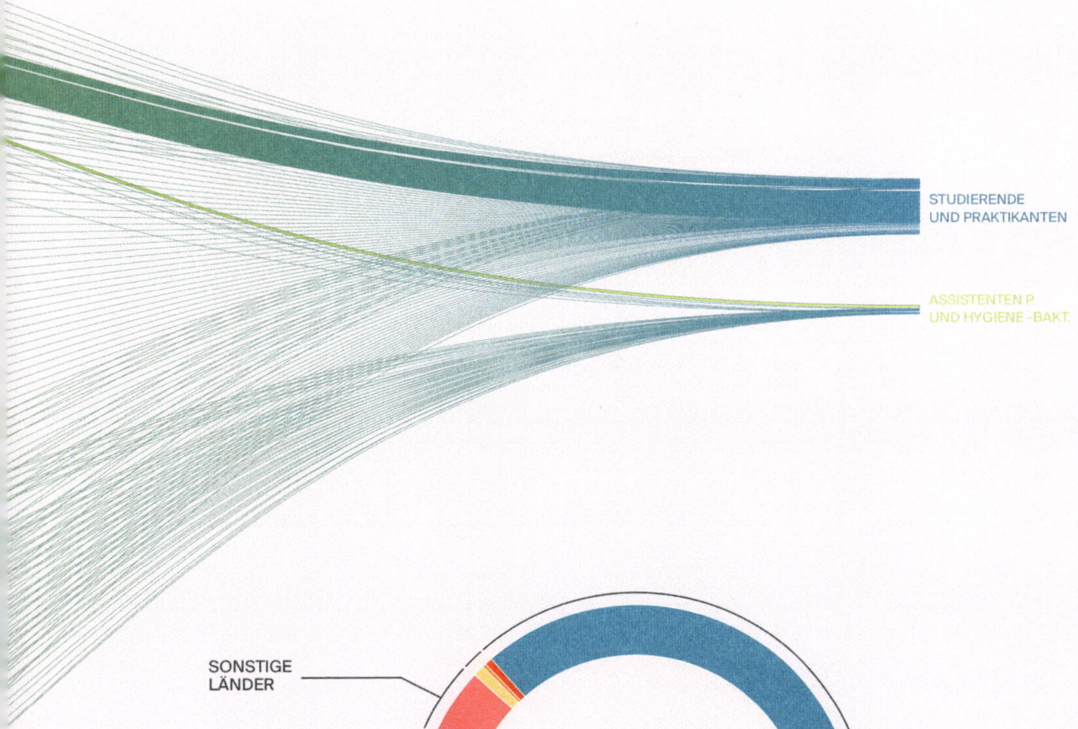
STUDIERENDE
UND PRAKTIKANTEN
ASSISTENTEN P
UND HYGIENE -BAKT.

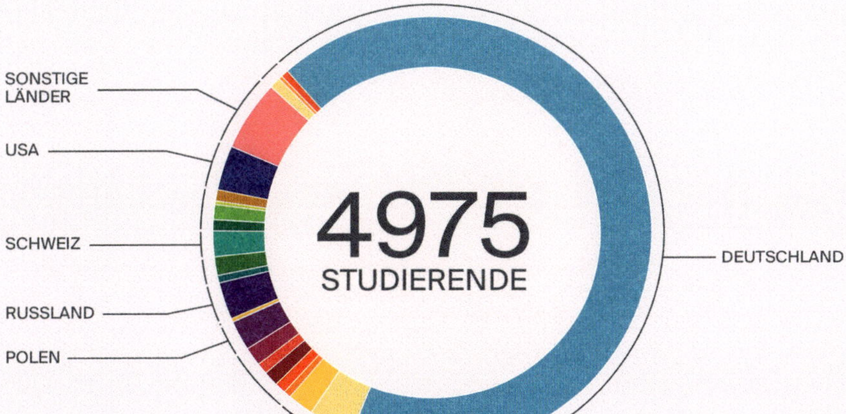
SONSTIGE
LÄNDER
USA
SCHWEIZ
RUSSLAND
POLEN
FRANKREICH
ENGLAND
4975
STUDIERENDE
DEUTSCHLAND

Grundriss des Chemischen Laboratoriums Fresenius (Erdgeschoss) von 1873[IV]

Floor plan of Fresenius' Chemical Laboratory, ground floor, 1873[IV]

Der Unternehmer

„... nicht unthätig zusehen, wenn das Unkräftig gewordene zusammensinkt ...“[33]

The entrepreneur

‘not to look on idly when what has lost its power begins to collapse ...’[33]

5

„Unternehmer sind keine Unterlasser!“ Auf Remigius Fresenius passt dieser Satz seines Ur-Urenkels Ludwig Fresenius, heute Ehrenpräsident der gleichnamigen Hochschule, sehr gut.

In einem Schreiben vom 7. April 1848 an die nassauische Regierung begründet der Gründer des Chemischen Laboratoriums das Festhalten an seiner unternehmerischen Idee auch in politisch bewegten Zeiten damit, man solle „... nicht unthätig zusehen, wenn das Unkräftig gewordene zusammensinkt ...“[33] Trotz Genehmigungsschrei-

‘Entrepreneurs never stop being enterprising!’ This saying, coined by Ludwig Fresenius, today Honorary President of the university that bears his family name, applies very well to his great-grandfather Carl Remigius Fresenius.

In a letter to the government of Nassau dated 7 April 1848 – that is to say, at a time of political turbulence – the founder of the Chemical Laboratory argued in favour of remaining faithful to his entrepreneurial concept by stating that one should ‘not look on idly when ben vom 13. Januar und finanziellen Zusagen hatten nämlich die Märzrevolution von 1848 und die nachfolgende behördliche Umorganisation das Vorhaben des jungen Professors zum Stocken gebracht. Er setzte seine Planung trotzdem um und begann am 1. Mai 1848 mit Dr. Emil Erlenmayer als Assistent und fünf Schülern den Betrieb. Als 1852 die Gesellschaft Deutscher Naturforscher und Ärzte unter seiner Leitung in Wiesbaden tagte, konnte Fresenius die Teilnehmer durch sein erweitertes Laboratorium führen. Die Regierung stellte fest, dass sich „die Einrichtung ... als vorzüglich bewiesen" habe und sah in Unterricht und Dienstleistungsanalysen Nutzen für den Staat. In mehreren weiteren Schritten baute Remigius die Kapazitäten zielstrebig aus und trug dabei, trotz staatlicher Zuschüsse immer wieder das unternehmerische Risiko selbst. Wir können davon ausgehen, dass seine nicht unbeträchtlichen Einkünfte aus den erfolgreich vielfach aufgelegten Lehrbüchern, die Einnahmen aus Auftragsuntersuchungen und die Studiengebühren eine Mischfinanzierung bildeten, aus der er Investitionen, laufende Kosten, Unterhalt für sich und die Seinen und Zukunftssicherung zu finanzieren hatte. Dabei begann er den wissenschaftlichen und gesellschaftlichen Erfordernissen gemäß mehrfach Neues: 1863 die pharmazeutische Lehranstalt (der Preußen 4 Jahre später, ein herber Rückschlag, die volle Anerkennung verweigerte), 1868 die önologische Versuchsstation (die der Staat später übernahm und teilweise nach Geisenheim verlegte), 1874 die Ausbildung von Nahrungsmittelchemikern, 1884 die bakteriologische Abteilung (deren medizi-

what has lost its power begins to collapse.'[33] In spite of a notification of official permission of 13 January and pledges of financial support, the March Revolution and subsequent reorganization of the local authorities of 1848 had seriously slowed down the young professor's intended project. Nevertheless, he continued to implement his plans and on 1 May 1848 commenced operations with Dr Emil Erlenmayer as his assistant and five students. When in 1852 the Society of German Natural Historians and Physicians held their conference under Fresenius's directorship in Wiesbaden, he was able to take the participants on a guided tour of his extended laboratory. The government was convinced that 'the institution [had] ... proved its excellence' and saw it as useful to the state in the field of education and as a provider of analyses on a commercial basis. In several further steps, Remigius single-mindedly increased its capabilities and, in spite of receiving state subsidies, repeatedly bore the entrepreneurial risk himself. We may assume that the not inconsiderable earnings from the many editions of his successful textbooks, the income from investigations conducted for external clients, and the study fees combined to form a mixed fund upon which he drew in order to finance his investments, his running costs, his and his family's living expenses, and insurance policies and the like taken out to safeguard the future. On this basis he launched out upon several new initiatives in response to the scientific and social requirements of the time: in 1863 the pharmaceutical college (to which, in a

Galerie südlich des sogenannten Weinlabors im alten Chemischen Laboratorium. Sie diente dem Gründer als regensichere "Wandelhalle". Rechts im Bild an der Wand einige der Sinnsprüche des Gründers aus dem Poesiealbum.

Gallery south of the former wine laboratory of Chemisches Laboratorium Fresenius. For the founder, it was a rainproof colonnade. To the right, some of Fresenius' epigrams as wall decoration.

nisch-mikrobiologische Aufgaben ein Vierteljahrhundert später von staatlichen Laboratorien übernommen wurden).

Die Rede des 27-jährigen Remigius Fresenius auf der Generalversammlung des Gewerbevereins für das Herzogthum Nassau am 9. Dezember 1845[34] erwies sich als programmatisch. Er führte aus, dass es „dem Gewerbestande nicht wohl möglich sei, sich mit dem Gebiet der chemischen Wissenschaft gehörig bekannt zu machen“. Daher sei es den Unternehmen förderlicher, ausgebildete Chemiker zu Rate zu ziehen, wenn es sich bei ihren Fragestellungen eher um Stoff- als um bloße Formveränderung handle. Seine nachhaltigen Beiträge zur wirtschaftlichen Entwicklung der Region und weit darüber hinaus waren Ausbildung praxisfähiger Fachkräfte, Dienstleistung in einem breiten Bereich von praktischen Fragestellungen (» Kapitel 3) und Verbreitung neuer Erkenntnisse durch Lehrbücher und Zeitschriften. Wäre es im Hinblick auf all dies nicht angemessen, sein Laboratorium einen „incubator service“ zu nennen – eine Brutstätte, von der vielfältige Anregung und Förderung ausgingen?

Schließlich darf nicht unerwähnt bleiben, dass Fresenius mit seinem Neffen Heinrich Dietze in Lorch im Rheingau eine Fabrik zur Erzeugung von Essigsäure und ihren Salzen aus Holz aufbaute, aus der der Verein für Chemische Industrie Mainz entstand. In diesem, im Verein chemischer Fabriken Mannheim und der Heidelberger (später Schweinfurter) Ultramarin-Farbenfabrik war er Aufsichtsratsmitglied und Aktionär.

Die Hochschule, die aus seinem Laboratorium hervorging, ist eine für

bitter setback four years later, Prussia refused to accord full recognition), in 1868 the oenological experimental research station (which was later taken over by the state and partially transferred to Geisenheim), in 1874 training courses for food chemists, in 1884 the bacteriological department (whose medico-microbiological tasks were taken over only a quarter of a century later by state laboratories).

The speech that the 27-year-old Remigius Fresenius had delivered at the Association of Trade for the Duchy of Nassau on 9 December 1845[34] turned out to be prophetic. In it he had noted that ‘the trade and industry sector seemed incapable of acquiring a proper familiarity with the field of chemical science.’ Therefore, when enterprises were dealing with problems related to the transformation of matter and not simply of form, they were best advised to seek the advice of trained chemists. His most enduring contributions to the economic development both within the region and far beyond its borders were the following: the formation of specialist professionals capable of applying their training in practice, the provision of services related to a broad range of practical questions and problems (» chapter 3), and the propagation of new discoveries and knowledge through textbooks and periodicals. This being the case, might his laboratory not very fittingly be described as an ‘incubator service’ – a breeding ground that sent out a wide variety of impulses and stimuli to the world outside?

angewandte Wissenschaften und knüpft damit an die Tradition ihres Gründers an. 2018 bereitet sie den Aufbau eines Gründerzentrums vor, das Infrastruktur und Beratung für neue geschäftliche Ideen bereitstellt. Ob der Gründer Remigius Fresenius verstehen würde, was wir heute meinen, wenn wir von „start-ups“ sprechen? Vermutlich schon! ●

Finally, it should not be forgotten that Fresenius and his nephew Heinrich Dietze built up a factory at Lorch im Rheingau for the production of acetic acid and its salts from wood. This enterprise grew into the Mainz Association for Chemical Industry, in which Fresenius was a shareholder and member of the supervisory board, as he was in the Mannheim Association of Chemical Factories and the Heidelberg Ultramarine Paint Factory (later located at Schweinfurt).

The university that developed out of Fresenius's laboratory is dedicated to the applied sciences and thus stands firmly in its founder's best tradition. In 2018 it is preparing to create a 'Founder's Centre' designed to offer infrastructure and advice for the implementation of new business ideas. Would Carl Remigius Fresenius understand what we mean when we talk about 'start-ups'? Most probably he would! ●

Das Chemische Laboratorium in der Kapellenstraße. Nur das Gebäude rechts mit den beiden Kellerfenster-chen steht noch und ist im Familien-besitz. Rechts von der rechteckigen Torfahrt findet man heute die Bronze-tafel "Historische Stätte der Chemie". Postkarte von 1904.

The Chemical Laboratory at Kapellenstraße. Only the building with the two small cellar windows to the right still exists. To the right of the rectangular portal, the bronze plaque 'Historical Site of Chemistry.' Postcard from 1904.

Kapellenstrasse m. chem. Laboratorium Fresenius.

Gereimtes

Geht es Dir im Leben gut,
Wahre Dich vor Übermut.
Gehen schlechter die Geschäfte,
So verdopp'le Deine Kräfte.

In rhyme

Is your business running well,
On high spirits never dwell.
In a period of depression,
Your endeavours try to freshen.

Der Bürger

„Die Kunst verdient, geschützt zu werden ...“[35]

The citizen

‘Art deserves to be protected ...’[35]

6

„So faß und ergreife ich dich Nassauer Land“. Diesen Satz ruft der 27 Jahre junge Remi in seinem ersten Wiesbadener Jahr aus, als er im Dunkel der Taunusstraße über eine Kette fällt. Wo ist da der Witz? Der als Frankfurter Bürgersohn geborene Remigius war mit dem Beginn seiner Tätigkeit am Landwirtschaftlichen Institut Geisberg Wiesbadener und damit Nassauer Bürger geworden. Er konnte viele Gedichte auswendig und zitiert hier abgewandelt Ludwig Uhlands Verse über Wilhelm den Eroberer, der mit diesen Worten „Engelland

‘Thus I grasp and take hold of you, Nassovian land,’ exclaimed the 27-year-old Remi when he stumbled over a chain in the darkness of the Taunusstrasse during his first year in Wiesbaden. What was the joke? When Remigius, who had been born the son of a Frankfurter, began to work at the Geisberg Agricultural Institute, he became a Wiesbadener and thus a citizen of Nassau. He knew many poems by heart and what he came out with on the Taunusstrasse was a paraphrase on the lines that Ludwig Uhland put into the mouth of

Die Stadt Wiesbaden setzte ihrem verdienten Ehrenbürger Carl Remigius Fresenius 1904 ein Denkmal in den Dambachtalanlagen (am Anfang der nach ihm benannten Freseniusstraße). Das Denkmal ist ein Werk des Berliner Bildhauers Karl Reinert.

The city of Wiesbaden raised a monument for its honorary citizen in 1904. It is situated in the Dambachtalanlage (at the beginning of Freseniusstrasse, named after him). It is a work of the sculptor Karl Reinert from Berlin.

ergreift"[36]. Als Eroberer kam er natürlich nicht nach Wiesbaden: Fresenius begreift das Bürgersein als Verpflichtung, sich für das Gemeinwesen zu engagieren.

1847 wird er in die Nassauische Ständeversammlung gewählt, aus der er 1852 wegen der zunehmenden Arbeitsbelastung in seinem Laboratorium wieder ausscheidet. In diese Zeit fällt der sogenannte Theaterstreit von 1848 um die Finanzierung des Wiesbadener Theaters. Wie heutig erscheint uns der Sparvorschlag, der Förderung bedürfe es nicht; Theaterfreunde könnten ja nach Mainz oder Frankfurt ausweichen! Dem stellt sich Fresenius wortmächtig entgegen: „Es wäre wahrhaftig kein Fortschritt, sondern ein Rückschritt, wenn man wegen den materiellen Interessen die geistigen gänzlich in den Hintergrund drücken würde; die Kunst verdient, wie der Ackerbau, die Gewerbe, sie verdient wie die Wissenschaft geschützt zu werden."[35] Ein Jahr lang ist er sogar Mitglied der siebenköpfigen Bürgerkommission, die das Theater nach Ende der herzoglichen Intendanz leitet.

1848 gehört der nationalliberal gesinnte Bürger Fresenius zu den etwa 1000 Unterzeichnern einer parteiübergreifenden Petition an das Frankfurter Vorparlament, in der man die Bereitschaft bekundet, die Rechte des Volkes mit „Gut und Blut" zu verteidigen.[37] Andererseits mahnt er 1849 in einer Rede[38] zur Besonnenheit. Kommt uns da etwas in der vom Internet beförderten öffentlichen Aufgeregtheit unserer Tage bekannt vor, wenn er hinzufügt: „Es ist nicht allein schwieriger, besonnen zu handeln, sondern es gehört auch mehr Muth dazu ..." und weiter erklärt, der Besonnene werde in der Regel

William the Conqueror as he set foot upon 'Engelland'.[36] Fresenius, of course, did not come to Wiesbaden as a conqueror: on the contrary, he regarded citizenship as an obligation to become active in the cause of the common weal.

In 1847 he was elected to the Assembly of the Estates of Nassau, from which he resigned in 1852 on account of the increasing workload at his laboratory. This period witnessed the controversy of 1848 concerning the financing of the Wiesbaden theatre. The argument for cutting the theatre's subsidy sounds positively modern: financial support was unnecessary, it was stated, because theatre-lovers could always go to Mainz or Frankfurt instead! Fresenius responded with powerful words: 'It would be no progress but a veritable step backwards if material interests were to push cultural interests entirely into the background; art, like agriculture or manufacturing and commerce, deserves to be protected, just as much as science.'[35] For one year Fresenius was even a member of the seven-man commission that took over responsibility for the theatre following the demise of the ducal directorship.

In his political opinions, Fresenius was a national liberal. In 1848 he was one of around a thousand signatories of a cross-party petition to the preliminary parliament in Frankfurt proclaiming their readiness to defend the rights of the people with their 'property and blood'.[37] In 1849, on the other hand, we find him making a speech[38] calling for prudence, adding: 'To act prudently is not only more difficult, but also calls for more courage.' In all this, we can sense something of the over-excited public discourse promoted by als Verräter bezeichnet – ihm dagegen komme es allein auf die Sache an?

Viele Jahrzehnte lang ist Carl Remigius Fresenius Mitglied des Gemeinde-, später Stadtrats, ab 1891 bis zu seinem Tod als gewählter Stadtverordnetenvorsteher.

Bei einem Chemiker mag es nicht verwundern, dass er dem Nassauischen Verein für Naturkunde bereits 1846 beitrat und ihn, später zusammen mit seinen Mitarbeitern, jahrzehntelang mit Publikationen und Vorträgen bereicherte. 1864-74 war er Director des Vereins und wurde danach zum Ehrenmitglied ernannt. Sein von Jugend an auch den Geisteswissenschaften zugeneigtes Naturell ließ ihn nicht einseitig werden. 1847 finden wir ihn unter den Gründern des Kunstvereins. Er war auch Mitglied im Verein für Nassauische Altertumskunde und Geschichtsforschung.

Als Enkel des von Goethe gerühmten Pfarrers Johann Philipp Fresenius blieb der von seinem Vater religiös erzogene Wissenschaftler seiner evangelischen Kirche stets treu verbunden. Die Pfarrer-Grabredner sprechen von Bekenntnis, Gottesdienstbesuch und von der Mitgliedschaft in kirchlichen Verbänden wie dem Protestantenverein.

In § 11 seines Testamentes[39] vermacht Fresenius der Stadt Wiesbaden 1000 Mark „mit der Bitte, sie in geeigneter Weise für die Armen Wiesbadens zu verwenden".

Als Bürger, so darf man zusammenfassen, hat Carl Remigius Fresenius das Seine getan. ●

the Internet in our own day, especially when he continues by noting that the man of prudence is generally called a traitor – and that he himself, by contrast, was solely concerned with the matter in hand?

For many decades Carl Remigius Fresenius was a member of the Wiesbaden City Council; from 1891 until his death he held the elected post of dean of the city councillors.

It is not surprising that Fresenius, as a chemist, was a member of the Nassau Natural History Society, which he joined in 1846. Over the decades it benefited greatly from the writings and lectures of Fresenius and his colleagues; in the years 1864–1874 he was the society's director, and was then made an honorary member. The fact that from an early age he had a natural inclination towards the humanities prevented him from becoming one-sided. In 1847 he was amongst the founder members of the Art Association; he was a member of the Society of Nassovian Antiquities and Historical Research.

Grandson of the pastor Johann Philipp Fresenius so highly praised by Goethe, Remigius was brought up in the Christian faith by his father and remained a loyal protestant his whole life long. The pastors who paid tribute to him at his funeral spoke of his readiness to bear witness to his faith, his attendance at church, and his membership in church bodies such as the inter-denominational Protestants' Union.

In para 11 of his Will[39], Fresenius left a thousand marks to the city of Wiesbaden 'with the request that they should be used in some appropriate way for the poor of Wiesbaden'. As a citizen, to put it in a nutshell, Fresenius certainly did his bit. ●

Carl Remigius Fresenius, 1880

Das Erbe

„… sein Wissen und Können in den Dienst der Allgemeinheit gestellt“[40]

The heritage

„… rich in knowledge and skills with which he served the general public …“[40]

7

Im Hauptstaatsarchiv Wiesbaden findet man das handschriftliche Testament von Carl Remigius Fresenius.[39] In 14 Paragaphen auf 15 Seiten verfügt er am 8. Januar 1894 akribisch, was mit Immobilien, Inventar und Vermögen (darunter zahlreiche Aktien und Versicherungspapiere) nach seinem Tod geschehen soll. Aus dem Vermögen soll seiner zweiten Frau Auguste ein jährlicher Mindestbetrag von 5000 Mark zur Verfügung stehen (damals verdiente ein Chemiearbeiter etwa 1500 Mark im Jahr). Mehrfach bietet er seinen

The manuscript Will of Carl Remigius Fresenius is preserved at the State Archives in Wiesbaden.[39] In 14 paragraphs spread over 15 pages penned on 8 January 1894 he laid down with meticulous care what was to happen to his real estate, goods and chattels, and financial assets (including numerous shares and insurance papers) after his death. The assets were to provide his second wife Auguste with an annuity of at least 5000 marks (at that time, a chemical worker earned around 1500 marks per annum). In several

Erben unterschiedliche Optionen an, deren Umsetzung davon abhängt, ob sie das Laboratorium fortführen wollen oder nicht. War er sich da nicht sicher, oder wollte er sie frei entscheiden lassen? Seine Söhne Heinrich und Theodor Wilhelm sind bei seinem Tod 50 und 41 Jahre alt und arbeiten schon längst in der Leitung des Laboratoriums mit, sein dann 43-jähriger Schwiegersohn Ernst Hintz ebenfalls. Sie machen alle drei weiter, so wie die Enkel Ludwig und Wilhelm und die Urenkel Wilhelm, Birgit und Remigius. Mit ihnen endet die Tradition des reinen Familienunternehmens. Trotzdem ist es ein seltener Fall, dass die drei Hauptstränge der Aktivitäten des Gründers in den Bereichen Bildung und Forschung, Dienstleistung und Fachzeitschrift weiter bestehen. Institut und Hochschule führen auch heute die Traditionsmarke Fresenius im Namen, die Zeitschrift setzt die Nummerierung von 1862 an nahtlos fort. In allen drei Bereichen ging es darum, nachhaltige Lösungen für die Zukunft zu gestalten.

Das Dienstleistungslabor hatte als freiberufliche Aktivität eines Mannes und seiner Mitarbeiter begonnen, das war es auch bis in die vierte Generation mit Wilhelm Fresenius. In der fünften Generation unter Ludwig Fresenius entstand daraus ein Netzwerk, eine Firma, die nun unter dem Namen SGS-Institut Fresenius zur weltweit aktiven, 1878 gegründeten Société Générale de Surveillance mit 95 000 Mitarbeitern und mehr als 2400 Niederlassungen und Laboren in über 120 Ländern gehört . Am Standort Taunusstein von SGS-Institut Fresenius analysieren, testen und bewerten 650 Männer und

instances he offers his heirs various options, depending on whether or not they wished to continue running the laboratory. Was he not sure of their intentions? Or did he perhaps just want to give them the freedom to decide for themselves? At the time of his death, his sons Heinrich and Theodor Wilhelm were 50 and 41 years of age and had both worked in the directorate of the laboratory for many years, as had his 43-year-old son-in-law Ernst Hintz. They all continued in the founder's footsteps, as did his grandsons Ludwig and Wilhelm and the great-grandchildren Wilhelm, Birgit and Remigius, with whom the tradition of the pure family business came to an end. Nevertheless, the Fresenius story is a rare case in that the three principal strands of the founder's activity – education and research, services, and the journal – continued to flourish long after his death. Today, both the SGS-Institut and the University of Applied Sciences still bear the 'brand name' Fresenius, and the serial numbering of the journal has gone on uninterrupted since 1862. In all three fields, Fresenius deliberately made it a priority to seek sound solutions that would survive the passage of time.

The laboratory dedicated to the provision of external services had begun as the free-lance independent activity of one man and his staff, which is what it remained through to the fourth generation under Wilhelm Fresenius. Now, after having developed into a network company in the fifth generation under Ludwig Fresenius, it goes under the name SGS-Institut Fresenius, being part of the

Frauen Mineral- und Trinkwässer, Rohstoffe, Lebensmittel, Kosmetika und Gebrauchsgegenstände und technische Produkte.

Ab 1990 erschienen die Arbeiten der von da an „Fresenius' Journal of Analytical Chemistry" genannten Fachzeitschrift nur noch in englischer Sprache. Aus den jährlich etwa 500 DIN A 5-Seiten im 19. Jahrhundert sind im Jahr 2017 7200 im DIN A 4-nahen Format der heutigen Zeitschrift ABC (» Kapitel 2.3) geworden. In zusammen 410 Bänden stecken von der Gründung bis Juni 2018 gerechnet fast 2500 Hefte und knapp 100 000 Artikel. Die Autorinnen und Autoren sowie das Herausgeberteam sind über den ganzen Globus verteilt.

Die Trägerin der Bildungs- und Forschungsstätte ist die Hochschule Fresenius gGmbH. Sie ist heute Teil der Cognos-Gruppe, in deren Bildungsgängen 1850 Mitarbeiterinnen und Mitarbeiter derzeit 19 000 junge Menschen unterrichten und anleiten. Der Mehrheitsgesellschafter von Cognos ist der Ururenkel des Gründers und Ehrenpräsident der Hochschule, Ludwig Fresenius, der damit die Familientradition in der fünften Generation weiterführt. Außer der Hochschulbildung deckt Cognos in mit den Ludwig-Fresenius-Schulen auch den Bereich berufliche Bildung ab. Die berufliche Bildung in den Bereichen Chemie und Biologie ist dem Fachbereich der Hochschule angeschlossen.

Die Hochschule Fresenius hat ihre Angebote in den Jahren seit 1996 schrittweise erweitert und bietet nun 56 Bachelor- und 39 Masterstudiengänge an zusammen 9 Standorten in Deutschland sowie in New York an. Medizinnahe Berufe lernt man in der nach dem Gründer benannten Carl Remigius School.

Société Générale de Surveillance founded in 1878, which has a total of 95,000 staff working at 2,400 branches and laboratories in over 120 countries. 650 men and women of the SGS Institut Fresenius at the Taunusstein site conduct analyses, testing and evaluating mineral and drinking waters, raw materials, foodstuffs, cosmetics, household articles and technical products.

Since 1990 the journal has been known as Fresenius' Journal of Analytical Chemistry and is published exclusively in English. While the nineteenth-century journal occupied a total of around 500 DIN A5 pages per year, in 2017 the ABC (» chapter 2.3) ran to around 7200 near DIN A4-pages. The series from the foundation through to June 2018 fills 410 volumes containing almost 2500 issues and only a little under 100,000 articles. The authors and team of editors are spread across the entire world.

Nowadays, the education and research programme is operated by and at the Fresenius University of Applied Sciences, which is a private limited company in its own right and now part of the Cognos Group. In the courses run by the Cognos Group, the majority shareholder of which is Ludwig Fresenius, a staff of around 1850 teach and instruct around 19,000 young people. Ludwig Fresenius is Honorary President of the Fresenius University and represents the fifth generation to carry on the family tradition. In addition to offering higher education, Cognos also provides vocational training through the Ludwig Fresenius Schools. Vocational schools for chemistry and biology are part of the

Heinrich Fresenius (1847-1920), ältester Sohn und ein Nachfolger des Gründers, hält eine Experimentailvorlesung. Postkarte aus dem Familienarchiv, abgestempelt am 20.11.1907

Heinrich Fresenius (1847-1920), eldest son and one of the founder's successors, giving an experimental lecture. Postard from the family archive, stamped on 20 November, 1907

Ihren derzeit über 11 000 Studierenden möchte die Hochschule in der Tradition ihres Gründers Wissen und Können vermitteln – und dazu, ganz im Sinn von Carl Remigius Fresenius, das Quäntchen Leidenschaft oder Lust und Liebe, ohne die ein Studium nicht gelingen kann.

Dieser Katalog und die Ausstellung, zu der er erscheint, sind vor allem der traditionellen Kernkompetenz in den Bereichen Chemie, Biologie und Pharmazie gewidmet, die ein begeistertes Team im Fachbereich Chemie & Biologie am Standort Idstein 200 Jahre nach „Remis“ Geburt in Lehre, Forschung und Praxis mit 500 Studierenden, Schülerinnen und Schülern teilt. Über die Hälfte der Studierenden verbringt mindestens ein Semester im Ausland, viele erlernen eine zweite Fremdsprache nach Englisch, ein Master-Kurs wird in dieser lingua franca der Chemie angeboten. Kooperationen in der angewandten Forschung verbinden die Hochschule mit Partnern in der Region, in Europa und weltweit.

Würde Fresenius, dem ökonomische Fragen nicht fremd waren, den seit den 90er Jahren schrittweise entstandenen Studiengang Wirtschaftschemie gut finden? Wahrscheinlich ja! Hätte er sich vorstellen können, dass Zeugnisse mit dem Namen Fresenius einmal die Leistungen von Studierenden in den Fachbereichen Wirtschaft & Medien, Gesundheit & Soziales und Design bescheinigen? Für die „Erben“ lag es jedenfalls nahe, dass dieser Katalog von einem Design-Professor der Hochschule gestaltet wird, dass eine von einer seiner Kolleginnen modern gestaltete Sitzfigur des Gründers die Besucherinnen und Besucher der Ausstellung empfängt

according department of Hochschule Fresenius.

Since 1996, the Fresenius University has gradually extended the scope of its courses and now offers 56 courses leading to Bachelor's degrees and 39 Master's courses at a total of 9 locations in Germany and one in New York. Occupations in the field of medicine are taught in the Carl Remigius School named after the original founder.

It is the University's intention to pass on knowledge and skills to its 11,000+ students in accordance with the tradition of its founder and moreover – absolutely in the spirit of Carl Remigius Fresenius – to generate in them the critical measure of passion (or heart and soul) for their chosen subject that is necessary for real success in any course of study.

The present catalogue and the exhibition for which it has been produced are dedicated principally to the traditional core competencies in chemistry, biology and pharmacology that – 200 years after Remi's birth – are imparted to 500 students and pupils by a highly motivated team of teachers in the special field Chemistry & Biology at the location of Idstein, not only through teaching but also through research and practical work. Over half the students spend at least one semester abroad and many learn a second foreign language in addition to English; as English is the lingua franca of chemists worldwide, a Master's course is also offered in this language. The University engages in numerous applied research cooperation programmes with partners in the region, in Europe and worldwide.

und dass kurze anschauliche Filme von Studierenden und Professoren der Medienstudiengänge sie informieren.

Ist das alles noch Fresenius? Der Gründer hatte schon als 30-jähriger von „zeitgemäßem, festem Neubau"[33] gesprochen. Ihn ernst nehmen, heißt für die Erben also gewiss nicht: Stehen bleiben. ●

Would Fresenius, who was no foreigner to the nitty-gritty of economic matters, have approved of the degree course in Business Chemistry that has been developed step by step since the 1990s? We think he would. But could he ever have imagined degree certificates bearing the name Fresenius being issued to students graduating from faculties of Economics & Media, Health & Social, or Design? Whatever he might have thought, his 'heirs' have considered it appropriate that the present catalogue should be created by one of the professors from the faculty of Design, and that visitors to the exhibition should be greeted by a thoroughly modern sitting figure of the founder designed by one of his colleagues, and be offered information and stimulation through short and visually appealing films produced by students and professors of the media study courses.

Is all this still Fresenius? Even at the age of 30 our founding father spoke of the importance of 'new structures, sound and up-to-date'[33]. If we take him seriously, there is one thing the heirs should definitely not do: stand still. ●

HOCHSCHULE FRESENIUS

So sehen die Standorte Idstein der Hochschule Fresenius und Taunusstein von SGS Institut Fresenius heute aus.

This is what the Idstein (Hochschule Fresenius) and Taunusstein (SGS Institut Fresenius) sites look today.

Zeittafel zum Leben von Carl Remigius Fresenius[A,B]

Timeline to the life of Carl Remigius Fresenius[A,B]

1818

Geburt von Carl Remigius Fresenius am 28. Dezember als Enkel des Taufpfarrers von J.W. von Goethe, Johann Philipp Fresenius, und als Sohn des Advokaten Jacob Heinrich Samuel Fresenius und seiner Frau Marie Veronika Finger in Frankfurt am Main.

Birth of Carl Remigius Fresenius on 28th December as a grandson of the pastor who baptized J.W. von Goethe, Johann Philipp Fresenius, and the son of lawyer Jacob Heinrich Samuel Fresenius and his wife Marie Veronika Finger in Frankfurt am Main.

1836

Nach dem Besuch der Musterschule in Frankfurt und des Benderschen Instituts in Weinheim an der Bergstraße wird Fresenius Lehrling in die Stein'sche Apotheke in Frankfurt. Besuch von Vorlesungen im Senckenberg'schen Institut. Erste chemische Versuche im eigenen Laboratorium im elterlichen Gartenhaus.

After attending the Musterschule in Frankfurt and the Bendersches Institut in Weinheim an der Bergstrasse, Fresenius becomes an apprentice in the Stein'sche Apotheke in Frankfurt. Visits lectures at the Senckenberg Institute. First chemical tests in own laboratory in the parental garden house.

1840

Fresenius studiert Chemie und Naturwissenschaften an der Universität Bonn. Eigene Experimente im Privatlaboratorium des Apothekers Dr. Clamor Marquart. Ihm widmet er die erste Auflage der 1841 erschienen „Anleitung zur Qualitativen Analyse".

Fresenius studies chemistry and natural sciences at the University of Bonn. Own experiments in the private laboratory of the pharmacist Clamor Marquart. To him he devotes the first edition of the 1841 published 'Guide to Qualitative Analysis'.

1862

Gründung der Zeitschrift für Analytische Chemie. Eröffnung der pharmazeutischen Lehranstalt.

Foundation of the Journal of Analytical Chemistry. Opening of the pharmaceutical school.

1873

Tod von Charlotte Rumpf, mit der Fresenius drei Söhne und vier Töchter hatte. Im Jahr darauf Eheschließung mit Auguste Fritze.

Death of Charlotte Rumpf, with whom Fresenius had three sons and four daughters. The following year, marriage with Auguste Fritze.

1875

Beginn der Ausbildung von Nahrungsmittelchemikern.

Beginning of training of food chemists.

1842

Fresenius wird Staatsassistent am Universitätslaboratorium zu Gießen und promoviert bei Justus Liebig. 1843 venia legendi (Lehrbefugnis) als Privatdozent in Gießen.

Fresenius becomes State Assistant at the University Laboratory in Giessen and holds a doctorate from Justus Liebig. 1843 venia legendi (teaching authorization) as a Privatdozent in Gießen.

1845

Berufung zum Professor am Herzoglich Nassauischen Landwirtschaftlichen Institut. Heirat mit Charlotte Rumpf, Umzug nach Wiesbaden.

Appointed professor at the Ducal Agricultural Institute of Nassau. Marriage with Charlotte Rumpf, move to Wiesbaden.

1848

Gründung des Chemischen Laboratoriums Fresenius mit Unterrichtslabor und Untersuchungslabor.

Foundation of the Chemical Laboratory Fresenius with teaching laboratory and research laboratory.

1884

Eröffnung der hygienisch-bakteriologischen Abteilung.

Opening of the hygienic-bacteriological department.

1897

Am 10. Juni stirbt Carl Remigius Fresenius und wird auf dem Alten Friedhof in der Platter Straße in Wiesbaden begraben. Auf dem Grabdenkmal steht: „Ich will Dich segnen, und Du sollst ein Segen sein". (1 Mos 12,2)

On 10th June Carl Remigius Fresenius dies and is buried in the Old Cemetery in Platter Street in Wiesbaden. On the tomb monument we read: 'I will bless you and thou shalt be a blessing.' (Genesis 12:2)

Hätte er sich das träumen lassen?

1975 hatte das Institut Fresenius nach Taunusstein, 1995 Hochschule Fresenius nach Idstein umziehen müssen. Im inzwischen um den Traditionsstandort gewachsenen Wohngebiet war kein Platz für die notwendigen Erweiterungen. 2017/18 ist ein Neubau an der Moritzstraße in Wiesbaden entstanden, der für etwa 75 Lehrende und 1000 Studierende Platz bietet. Die meisten Namen der Studiengänge, die dort angeboten werden, würden den Gründer mindestens zu Rückfragen veranlassen: Was bitte studiert Ihr da? 3-D Design und Management?

Would he have dreamed that?

In 2013, the Gesellschaft Deutscher Chemiker awarded the last surviving building of the Chemische Laboratorium Fresenius as one of 18 places in Germany with a bronze plaque as a historical site of chemistry. It annually awards the Fresenius Prize for Analytical Chemistry. The American Chemical Society (ACS) has the Phi Lambda Ypsilon Company. The Greek letters stand for the names Fresenius, Liebig and Van't Hoff. The ACS awards the National Fresenius Award to outstanding junior scientists, of whom Roald Hoffmann and Martin Karplus later received Nobel Prizes.

Eng verbündet mit Raum und Zeit...

Nein, Heldenverehrung wollen die Ausstellung anlässlich des 200. Geburtstages von Carl Remigius Fresenius und die vorliegende Schrift nicht betreiben. Die gezeigten historischen Geräte, der Rückblick in das vorvorige Jahrhundert und das Zitieren von Worten des Meisters und seiner Zeitgenossen sind nicht Reliquienpflege. Sie sind vielmehr der Versuch seiner „Erben", ihren Wurzeln nachzuspüren und dabei etwas über den Weg zu lernen, der bis zu ihnen in unsere Zeit führt.

War „Remi" nie enttäuscht? Manchmal fürchterlich. Konnte er nicht wenigstens ab und zu böse oder ärgerlich werden? Hoffentlich. Ist ihm nicht wenigstens das eine oder andere misslungen? Sicherlich. Er war, wie er selbst in seinem letzten Gedicht erkennt und wie es für uns alle gilt, ein Mensch in seiner Zeit. Wir haben uns ihm aus 7 Blickwinkeln tastend genähert. Dabei galt es, auszuwählen, wegzulassen, zuzuspitzen, zu vereinfachen.

Was kann heute und in Zukunft gelten, das „Remi" Richtschnur war? Er war auch deshalb ein Meister, weil er sich beschränkte auf das, was er besonders gut konnte. Und das versuchte er, besonders gut zu machen. Im heutigen Management-Deutsch: Er war effektiv und effizient. Er tat das, wozu er Lust und Liebe hatte, und er tat es mit Lust und Liebe. Er knüpfte Netzwerke, teilte Wissen und gab Wissen und Können weiter. Die „Modulbeschreibungen" unserer Studiengänge sprechen von knowledge und skills. Er eignete sich neue Erkenntnisse umgehend an, häufig nicht ohne sie selbst experimentell zu überprüfen. Wenn sie sich bewährten, wendete er sie praktisch an und teilte das in seiner Zeitschrift und seinen Büchern mit. In seinen praktischen Tätigkeiten achtete er auf das, was wir heute Nachhaltigkeit nennen, scheute sich aber auch nicht, das zurückzulassen, was nicht mehr taugte.

„Die Welt verändert sich rasend schnell. Wir müssen versuchen, nicht hinterher, sondern voranzuschreiten." Das sagte sein Urenkel Wilhelm Friedrich Nils Remigius an seinem 90. Geburtstag 2003. Das Voranschreiten muss die Aspekte der Wissenschaft, Praxis und Unternehmerisches im Blick behalten. In diesem Sinn freut sich sein Ur-Urenkel Ludwig schon auf das 175. Gründungsjahr des Laboratoriums 2023.

Ein George Bernard Shaw zugeschriebener[42] erleuchteter Satz sagt uns: „Tradition ist eine Laterne: Der dumme hält sich an ihr fest, dem Klugen leuchtet sie auf dem Weg." In diesem Sinn mag uns ein Licht aufgehen, wenn wir uns mit „Remi" beschäftigen. ●

Das meinen Andere

„Eine Tradition haben heißt nicht, eine Last seelenloser Bräuche und eines toten Wissens mit sich herumzuschleppen. Tradition heißt, hineintreten dürfen in den Strom, der sich nährt von den Wünschen und den Hoffnungen von vielen. In diesem Strom habe ich ein Grundwissen: Ich bin nicht allein, ich fange nicht an, ich muss mich nicht mit mir selber begnügen, nicht mit der eigenen Sprache und nicht mit dem eigenen kärglichen Geist. In unser Leben ist hineingewoben die Kraft und die Hoffnung all derer, die vor uns waren. Wir bauen unsere Lebenshäuser auf der Stärke und auf den Visionen der Toten. Wir haben Träume geerbt und so träumen wir nicht allein. Die Träume unserer Väter und Mütter sind nicht ausgeträumt, wir träumen sie weiter. Auch wir werden sie nicht zu Ende träumen. Wir hoffen auf unsere Kinder und Enkelkinder. Vielleicht kommen sie weiter als wir, vielleicht sind sie stärker, vielleicht sind sie mutiger. Man kann in eine neue Zeit treten, wenn man das Glück der eigenen Endlichkeit schätzt: Wir fangen nicht an, und wir müssen nicht vollenden.“[41]

What others think

‘Having a tradition does not mean lugging a load of soulless customs and dead knowledge with you. Tradition means to be allowed to step into the stream that feeds on the wishes and hopes of many. In this stream I have a basic knowledge: I am not alone, I do not start, I do not have to be content with myself, not with my own language and not with my own scanty mind. The power and hope of all who came before us has been woven into our lives. We build our homes on the strength and visions of the dead. We have inherited dreams and so we do not dream alone. The dreams of our fathers and mothers are not over, we dream them on. We too will not dream them to the end. We hope for our children and grandchildren. Maybe they will come further than us, maybe they are stronger, maybe they are more courageous. You can step into a new era, when you appreciate the happiness of your own finiteness: we do not start, and we do not have to complete.’[41]

Closely allied with time and space…

No, hero-worship is emphatically not the aim of the Carl Remigius Fresenius 200th anniversary exhibition or of the present booklet. Nor are the historical instruments on display intended as relics to be venerated – nor are the words of the Master and his contemporaries that we quote or our glance back at the last century but one. On the contrary, they represent an attempt on the part of his 'heirs' to go back to their roots and find out more about the path leading from those roots through to our own times.

Was 'Remi' never disappointed? Of course he was, sometimes terribly disappointed. Was he not capable of being offended or angry? We hope so! Didn't he too experience failure on occasion? Of course he did. As he himself recognized in his last poem, he was a man of his time, just as we too are all men and women of our time. We have tentatively approached him from seven points of view. In so doing we had to be selective, leaving some things out, highlighting certain aspects and simplifying others.

Of all Remi's guiding principles, which ones are still valid today and will remain so in the future? One of the things that made him a master was that he restricted himself to the things he was really good at and strove to do these things particularly well. In modern management terms: He was effective and efficient. He did what he genuinely wanted to do and did it with great enthusiasm. He built up networks and enjoyed sharing and imparting what the 'module descriptions' of our study courses call 'knowledge, skills and competencies'. When new discoveries and insights appeared, he immediately made them his own, frequently checking them by conducting experiments. When they stood the test, he disseminated them in his journal and his books. In all his undertakings he took account of what we today call 'sustainability', but he did not shrink from jettisoning what had lost its usefulness.

'The world is changing at breakneck speed. We must endeavour not to fall behind but to stride ahead.' These words were spoken by Remi's great-grandson Wilhelm Friedrich Nils Remigius on his 90th birthday in 2003. In 'striding ahead' we must keep our eyes on the three aspects of science, practice and enterprise. Accordingly, Remi's great-great-grandson Ludwig is already looking forward to 2023 and the 175th anniversary of the laboratory's foundation.

A sentence attributed to[42] George Bernard Shaw tells us: 'Tradition is a lantern: the foolish hold on to it, but the clever allow it to lead the way.' In this spirit, may a lantern shine brightly for us as we continue to study the life and work of our founder 'Remi'. ●

Bislang sind zwei umfassende Monographien zu Leben und Wirken von C.R. Fresenius erschienen. Beide enthalten umfassende Listen der Publikationen von Fresenius und seinen Mitarbeitern.

So far, two comprehensive monographs on C.R. Fresenius's life and achievements have been published.
They contain full lists of the publications of Fresenius and his collaborators.

A.
Susanne Poth, Carl Remigius Fresenius (1818-1897). Wegbereiter der analytischen Chemie. Heidelberger Schriften zur Pharmazie- und Naturwissenschaftsgeschichte. Stuttgart 2007.

B.
Walter Czysz, 140 Jahre Chemisches Laboratorium Fresenius Wiesbaden. 1. Teil: 1848-1945
Jb. Nass. Ver. Naturk. 110, 35-110 (1988)

Weitere wichtige Quellen – Further important sources

I.
R. Fresenius, Karl Remigius Fresenius 1818 bis 1898. In: R. Vaupel (Hrsg.), Nassauische Lebensbilder Band I. Veröffentlichungen der Historischen Kommission für Nassau X,1. Wiesbaden 1940, S. 191-208

II.
Wilhelm Fresenius, Zur Erinnerung an Remigius Fresenius. Erinnerungen aus meines Vaters jungen Jahren. Privatdruck o.J.

III.
Remigius Fresenius. Zur hundertsten Wiederkehr seines Geburtstages, 28. Dezember 1918. Chem. Ber. A 52 (2) 1919, S. 37-40

IV.
R. Fresenius, Geschichte des chemischen Laboratoriums zu Wiesbaden während der ersten 25 Jahre des Bestehens der Anstalt. Wiesbaden 1873

V.
H. Fresenius, Geschichte des chemischen Laboratoriums zu Wiesbaden während der zweiten 25 Jahre seines Bestehens. Wiesbaden 1898. S. VII f

VI.
H. Fresenius, Zur Erinnerung an R. Fresenius. Fres. Z. Anal. Chem. 36, II-XVIII (1897)

Anmerkungen/Literatur - Explanatory notes and sources:

1.
Gute Überblicke geben die Monographien A und B und die Quellen I-VI –
Good reviews are found in monographs A and B and the sources I-VI

2.
Quelle I, S. 202

3.
So sieht Fresenius den Wissenschaftler, der Neues sucht und findet. Aus einem handschriftlichen Manuskript von Carl Remigius Fresenius, durch Vergleich mit Briefen zwischen ihm und seinem Vater identifizierbar als Text für seinen Vortrag vor dem Senat der Universität Gießen zum Erhalt der akademischen Lehrbefugnis (venia legendi) am 27.01.1844.
This is how Fresenius sees the scientist seeking and gaining new findings. Taken from a manuscript which, through comparison with letters he exchanged with his father, can be identified as the text for the lecture he delivered to the senate of the University of Gießen on 27 January 1844 in order to obtain the licence to teach as a professor.

4.
Quelle II, S. 4

5.
Quelle I, S. 193

6.
Quelle II, S. 7

7.
C.R. Fresenius, Über das Thun und Treiben im chemischen Laboratorium zu Giessen, mit besonderer Berücksichtigung der Ergebnisse des letzten Jahres. In: Amtlicher Bericht der Gesellschaft Deutscher Naturforscher und Ärzte 1842 (1843), 92-110. Der Text wurde unter Korrektur von Lesefehlern der online-Quelle http://archive.org/stream/amtlicherbericht20gese/amtlicherbericht20gese_djvu.txt entnommen.
This text was taken from the online source cited, with errors made in transcription being corrected.

8.
C. R. Fresenius, Anleitung zur Quantitativen Chemischen Analyse, 6. Auflage 1875, Erster Band, S. 4.
English version taken from William H. Brock, The Fontana History of Chemistry. London 1992, S. 208, who uses the term „Hippocratic oath". Brock verwendet für diesen Text den Begriff „Hippokratischer Eid".

9.
J.A. Perez-Bustamente, Karl Remigius Fresenius (1818-1807):
Rationalization and systematization of chemical analysis. Química Analítica 16, 139-141 (1997)

10.
Ankündigung zum Erscheinen der Zeitschrift für Analytische Chemie 1861. Quelle VI, S. XIII und S. IX

11.
R. Fresenius, Chemische Untersuchung der wichtigsten Mineralwasser

des Herzogthums Nassau, Wiesbaden 1850, S. 1. Fresenius weist hier darauf hin, dass die Kenntnis der chemischen Zusammensetzung eines Mineralwassers nicht rein wissenschaftlichem Interesse dienen, sondern auch dem Brunnenbesitzer den materiellen Wert und dem Arzt den medizinischen Nutzen einschätzen helfen solle.

Here Fresenius points out that establishing the chemical composition of a mineral water does not only have scientific goals. It is also done to help the owner of the source estimate its economic value, and the physician to know its medical usefulness.

12.
Quelle III, S. 37

13.
Bericht der Handelskammer zum 25. Jubiläum des Chemischen Laboratoriums Fresenius, Quelle V, S.VII

14.
G. Schwedt, C. Remigius Fresenius und seine Mineralwasseranalysen. An den Quellen in und am Taunus. Aachen 2013. ISBN 978-3-95631-003-4. Auf den Seiten 233-236 findet sich eine umfassende Liste von 92 in Deutschland von Fresenius untersuchten Quellen mit Angabe der Fundstelle des Berichts.

On pages 233-36 there is a comprehensive list of 92 mineral water sources in Germany analysed by Fresenius, with the respective citations. - L. Gros und G. Schwedt, Carl Remigius Fresenius – ein Wegbereiter der Wasseranalytik im 19. Jahrhundert und seine Nachfolger in den ersten hundert Jahren. Vom Wasser 116 (1), 13-20 (2018)

15.
R. Fresenius, Ueber die Stellung des Chemikers bei gerichtlichen Untersuchungen und über die Anforderungen, welche von Seiten des Richters an ihn gemacht werden können. Annalen der Chemie und Pharmacie 49, 275 (1844).

16.
R. Fresenius, Beiträge zur gerichtlichen Chemie. Fresenius' Z. Anal. Chem. 6, 197 (1867)

17.
R. Fresenius, Lambert Babo, Ueber ein neues, unter allen Umständen sicheres Verfahren zur Ausmittelung und quantitativen Bestimmung des Arsens bei Vergiftungsfällen. Annalen der Chemie und Pharmacie 49, 287 (1844)

18.
R. Fresenius, Der Korneuburger Vergiftungs-Prozess, (1857-1859). Wien 1860. Hochschul- und Landesbibliothek 60 Nl 1837

19.
W. Fresenius (Sohn von/son of Remigius Fresenius), Welche Grundsätze sollen bei der chemischen Untersuchung und Begutachtung der Weine allgemein als Richtschnur dienen? Internationaler land- und forstwirtschaftlicher Congress zu Wien 1890. Section V: Landwirtschaftliches Unterrichts- und Versuchswesen. Subsection c, Frage 99, Gruppe C.a. Wien 1890, S. 3.

20.
Zwei wesentliche Werke sind: E. Borgmann, Anleitung zur chemischen Analyse des Weines. Wiesbaden 1884 sowie C. Neubauer, Über die Chemie des Weines. Drei Vorträge gehalten im Winter 1869/70 in Mainz, Oppenheim und Oestrich a.Rh. Wiesbaden 1870

21.
Die Wandmalereien im Ratskeller zu Wiesbaden. 124 Abbildungen nach den Originalhandzeichnungen von K. Kögler und H. Schlitt mit begleitendem Text. Wiesbaden o.J. (um 1905), S. 56 f. Die hier gezeigten Abbildungen und Texte waren im als Weinlokal dekorierten Keller angebracht.

The paintings and texts shown here ornamented the cellar decorated as a wine pub.

22.
R. Fresenius in einem Gutachten zur Zusammensetzung einer Schwefelkies-Probe aus Annerod bei Gießen. 17.05.1855. Familienarchiv Fresenius.

Fresenius in an expert report about the composition of an iron pyrite sample from Annerod near Gießen.

23.
R. Fresenius, Ueber die Ausfüllung des Mantels der feuerfesten Kassenschränke. Mittheilungen für den Gewerbeverein des Herzogthums Nassau 1847 (2), 1

24a.
R. Fresenius, W. Fresenius, Untersuchungen über den Nachweis von Verfälschungen im Portland-Cement. Fres. Z. Anal. Chem. 23, 175 (1885)

24b.
R. Fresenius, W. Fresenius, Fres. Z. Anal. Chem. 23,66 (1886)

24c.
R. Fresenius, W. Fresenius, Fres. Z. Anal. Chem. 32,433 (1893)

25.
Ernst Hintz, Über Gasglühlicht. Vortrag, gehalten am 25. Mai 1898 bei Gelegenheit der Feier des fünfzigjährigen Jubiläums des Chemischen Laboratoriums Fresenius zu Wiesbaden: nebst wissenschaftlichen Anlagen / von Ernst Hintz, Professor, Director und Mitinhaber des Chemischen Laboratoriums Fresenius zu Wiesbaden. Wiesbaden 1898. Hochschul- und Landesbibliothek Wiesbaden Vh 4431

26.
75 Jahre bakteriologische Abteilung des Chemischen Laboratoriums Fresenius Wiesbaden. Privatdruck o.J.

27.
C. Neubauer und J. Vogel, Anleitung zur qualitativen und quantitativen Analyse des Harns, sowie zur Beurtheilung der Veränderungen dieses Secrets mit besonderer Rücksicht auf die Zwecke des praktischen Arztes. Zum Gebrauche für Mediciner und Pharmazeuten. 2. Auflage 1856. Das Werk erschien in zahlreichen Neuauflagen bis ins 20. Jahrhundert.

28a.
F. Hüppe, Die Methoden der Bakterien-Forschung. Wiesbaden 1885 (weitere Auflagen folgten; followed by more editions).

28b.
Referat von F. Hüppe in: Bericht für den VI. internationalen Congress für Hygiene und Demographie zu Wien 1887. Heft 2, 3-19.

29.
Quelle III, S. 37

30.
Quelle IV, S. 16

31.
Quelle III, S. 37

32.
Mallinckrodt 125th Year Anniversary. 1867-1992. Quality, Integrity, Service. Written and produced by the History Factory. ISBN 1-882771-00-1. S. 9 f.

33.
Quelle IV, S. 7

34.
Bericht über die am 9. Dezember 1845 abgehaltene Generalversammlung des Gewerbevereins für das Herzogthum Nassau. Wiesbaden 1846. Hessische Hochschul- und Landesbibliothek Signatur Gv7

35.
Landständische Verhandlungen 1848. 32te Verhandlung am 24. August 1848, S. 106

36.
Quelle II, S. 9

37.
Thomas Weichel, Die Bürger von Wiesbaden. Von der Landstadt zur „Weltkurstadt“ 1780-1914. München 1997 S. 207

38.
Landständische Verhandlungen 1849. Verhandlung vom 15. Mai. S. 451

39.
Hauptstaatsarchiv Wiesbaden 469-33

40.
Quelle III, S. 40

41.
Fulbert Steffensky, Mit jeder Erinnerung geben wir uns ein Gesicht. PUBLIK-FORUM Nr. 4, 2003, S. 49

42.
siehe/see www.falschzitate.blogspot.com geprüft am/checked on 15.07.2018

43.
Letztes Gedicht von Remigius Fresenius, verfasst am Vorabend seines Todes.

Remigius Fresenius‘ last poem, written down in the evening before his death. Quelle I, S. 202 f.

Bildnachweise
Picture Credits

Anonymus, ca. 1843: Seiten 23, 115 (gemeinfrei)

Anonymus, ca. 1863: Seite 81 (gemeinfrei)

Archiv der Hochschule Fresenius, 2018: Seiten 96-97

Robert W. Bunsen und Gustav R. Kirchhoff, Zeitschrift für Analytische Chemie 1, 1862: Seite 34-35, 114 (gemeinfrei)

Evonik Industries AG, Konzernarchiv Hanau, 2018: Seite 114

Familienarchiv Fresenius, 2018: Seiten 2-3, 20, 52-53, 93, 108-109, 114

Bernd Fickert, Museum Wiesbaden, 2018: Seiten 10, 28-29, 31, 32, 33, 38, 39, 41, 42-43, 44, 45, 50, 59, 63, 77, 115. Alle Objekte außer S. 50 sind Leihgaben aus der Sammlung zur Laborgerätetechnik 1750 – 1950, Rainer Friedrich, Blankenfelde-Mahlow bei Berlin

Remigius Fresenius, Zeitschrift für Analytische Chemie 25, 26 (1886): S. 20; Anleitung zur qualitativen Analyse 11. Auflage 1862 S. 29: S. 35; Zeitschrift für Analytische Chemie 1 (1862): S. 48 (gemeinfrei)

Remigius Fresenius, Geschichte des chemischen Laboratoriums zu Wiesbaden während der ersten 25 Jahre des Bestehens der Anstalt, 1873: Seite 90 (gemeinfrei)

Remigius Fresenius und Wilhelm Fresenius, Zeitschrift für Analytische Chemie 23, 1884: Seite 73 (gemeinfrei)

Remigius Fresenius und Heinrich Will, New methods of alkalimetry and of determining the commercial value of acids and manganese, London 1843; sowie dieselben, Guide pratique pour reconnaître et pour déterminer le titre véritable et la valeur commerciale des cendres, des acides et des manganèses, Paris, 1844, Seite 70 (gemeinfrei)

Leo Gros, Hochschule Fresenius, 2018: Seite 84

Gesellschaft Deutscher Chemiker, 2013: Seite 127

Ulrich Henke, 2018: Seiten 16-17, 115 (mit freundlicher Genehmigung des Urhebers)

Hessisches Hauptstaatsarchiv Wiesbaden, Nr. 3008/1, 13029: Seite 104

Hochschule Fresenius, 2018: Seiten 112, 113

Katharina Schwochow, Infographics group Berlin und Dieter Dreher, Hochschule Fresenius, 2018: Seiten 88-89

Kaspar Kögler und Heinrich Schlitt, Die Wandmalereien im Ratskeller zu Wiesbaden, 1905: Seiten 66-67 (gemeinfrei)

Ludwig Knaus, Portrait von Fresenius, 1859: Seite 115, Museum Wiesbaden

Stadtarchiv Wiesbaden, Nr. 000805 Digitales Multimedia-Archiv: Seiten 100-101

Stadtarchiv Wiesbaden, Nr. 013984 Digitales Multimedia-Archiv: Seiten 82-83

SGS Institut Fresenius, 2018: Seite 113

Springer Verlag, 2018: Seite 48

Wilhelm Trautschold, Illustration in: "Bibliothek des allgemeinen und praktischen Wissens", Band 3, Deutsches Verlagshaus Bong u. Co., Berlin, 1912, Herausgeber Emanuel Müller-Baden: Seite 15 (gemeinfrei)

Clemens Winkler, Zeitschrift für Analytische Chemie 25, 1886: Seite 20 (gemeinfrei)

Die Beteiligten der Ausstellung
The participants of the Exhibition

Kuratoren
Curators
Leo Gros
Rüdiger Wandke
Fritz Geller-Grimm
Hannes Lerp
Susanne Kridlo

Gestaltung
Design
Fritz Geller-Grimm
Leo Gros
Hannes Lerp
Ann-Kathrin Kempnich

Aufbau
Installation
Felix Richter
Malte Seehausen
Gerhard Heinrich
Ann-Kathrin Kempnich
Rüdiger Wandke
Michael Edler
Michael Krag
Christian Rücker
Johanna Zwick

Technik
Technical maintenance
Michael Edler

Präparation
Preparation
Malte Seehausen
Felix Richter

Texte
Texts
Leo Gros
Hannes Lerp

Übersetzungen
Translations
John Nicholson
Leo Gros

Fotografie
Photography
Bernd Fickert

Grafik
Graphics
Bernd Fickert
Fritz Geller-Grimm
Ann-Kathrin Kempnich

Begleitprogramm und Registratur
Accompanying programme and registry
Susanne Kridlo
Gabriele Knepper

Öffentlichkeitsarbeit
Public relations
Susanne Löffler
Susanne Kridlo
Antonia Junker
Martina Brand

Administration
Administration
Jessica Krimme
Angelika Hermersdorfer
Heidi Piechnik

Aufsicht
Supervision
Torsten-Marius Röhnke-Tartler,
Stephan Müller
Adelheid Geisler
und die Kolleginnen und Kollegen der Firma Kötter

Leihgeber
Lenders
Rainer Friedrich, Blankenfelde-Mahlow bei Berlin, Sammlung zur Laborgerätetechnik 1750 – 1950
Ludwig und Margarethe Fresenius, Chur (CH), Berthold Fresenius (Frankfurt)
Renate Bress geb. Fresenius (+), Wiesbaden
Hochschule Fresenius, Idstein
Merck Corporate History Darmstadt
Selterswassermuseum Niederselters, Dr. Norbert Zabel
Hochschul- und Landesbibliothek Wiesbaden
Hauptstaatsarchiv Wiesbaden
Deutsche Metrohm GmbH, Filderstadt
Process [.-ING] GmbH, Frankfurt am Main
Diasys Diagnostic Systems GmbH, Holzheim
SGS Institut Fresenius GmbH, Hamburg; Standorte Taunusstein und München
Kiwa GmbH Servicecenter Flörsheim-Wicker
Weinlabor Vogel, Geisenheim

Förderer
Supporters
Hessen Agentur GmbH
SGS Institut Fresenius
HessenChemie e.V.
Alfred Weigle Stiftung Wiesbaden
Lions Club Wiesbaden
Naspa Nassauische Sparkasse
Nassauischer Verein für Naturkunde
Freunde des Museums Wiesbaden
hr2 kulturpartner

Danksagung – Der Autor dankt herzlich für Rat, Unterstützung und Hinweise
Acknowledgement – To the following, the author extends warmest thanks for advice, support and information.

Dr. Susanne Poth, Wiesbaden
Prof. Dr. Georg Schwedt, Bonn
Rainer Friedrich, Blankenfelde-Mahlow
Dr. Till Macke, Bonn
Jürgen Funk, HessenChemie
Dr. Gabriele Fünfrock, Wiesbaden
Dr. Klaus Eiler, Dr. Carl Christian Wahrmann, Dr. Hartmut Heinemann, Hauptstaatsarchiv Wiesbaden
Dr. Brigitte Streich, Stadtarchiv Wiesbaden
Dr. Martin Mayer, Hochschul- und Landesbibliothek Wiesbaden
Dr. Sabine Bernschneider-Reif, Archiv Merck
Dr. Thomas Geelhaar, Merck KGaA
Dr. Christa Jansen, Merck KGaA
Dr. Sabine Eiers, Merck KGaA
Dr. Andrea Hohmeyer, Evonik-Konzernarchiv
Dr. Ulrich Eisenbach, Hessisches Wirtschaftsarchiv
Mag. Wilma Buchinger, Österreichische Nationalbibliothek Wien
Dr. Susanne Rehn, Deutsches Museum München
Dr. Norbert Zabel und Eugen Caspary, Selters
Ursula von Breitenbach, Weingut Geromont Winkel
Prof. Ernő Keszei, Budapest
Prof. Francesco de Angelis, Acquila/Roma
Prof. Miquel Gassiot, Universitat Ramon Llull, Barcelona
Dieter Mallmann, Traben-Trarbach und Christel Fritsch, Rhaunen
Martin Möllmann, Dyckerhoff GmbH
Dr. Andreas Brockmann, Torsten Laub, Jan Deissner, Sebastian Rau, Tilman Heinisch, Dr. Jochen Dobberstein, SGS Institut Fresenius
Prof. Dr. Thorsten Daubenfeld, Prof. Dr. Thomas Knepper, Prof. Dr. Achim Menges, Sascha Kappes, Merve Preuß, Marion Schild, Hochschule Fresenius

Mitwirkende aus der Hochschule Fresenius

Prof. Manuel Trüdinger (Katalog)
Dieter Dreher
(mit Katharina Schwochow, Infographics Group Berlin)
(Graphik Studierende S. 88/89)
Prof. Ulrike Nägele
(Sitzfigur C.R. Fresenius)
Prof. Rob Negelen
Nicola Schott (Video Studierende Fresenius)
Nils Hachmeister (Videos Analyseverfahren und Hochschule Fresenius)

Rosmarie Daser-Martin
Dipl.-Chem. Wolfgang Noll
Prof. Dr. Leo Gros (Experimente Analysenverfahren)

Arne Ramstätter BSc und die Studierenden Anika Müller
Pia Neubert
Charlotte Needham
Tankred Lehmkuhl (Erstellung der Datenbanken zu Studierenden 1848-1948 und zu Literatur von und über Carl Remigius Fresenius)

Die Alumni Dipl.-Ing. Claudia Schultheiß, Dipl.-Ing. Bernd Geis und Dipl.-Ing. Nicole Machill

Impressum
Impress

Autor
Author
Leo Gros

Herausgeber
Editor
Museum Wiesbaden
Hochschule Fresenius

Redaktion und Lektorat
Editing and Copyediting
Hannes Lerp
Leo Gros
Fritz Geller-Grimm

Übersetzungen
Translations
John Nicholson
Leo Gros (Gedichte)

Korrektur
Copyediting
Hannes Lerp

Gestaltung
Graphic Design
Manuel Trüdinger
Felix Flemmer
Abc&D, München

Umschlagbilder
Cover Illustrations
Glasgefäße und Korken aus der Zeit Carl Remigius Fresenius'. Sammlung Friedrich. Foto: Bernd Fickert, Museum Wiesbaden

Bibliografische Informationen
Die Deutsche Bibliothek verzeichnet diese Publikation in der Deutschen Nationalbibliografie; detaillierte bibliografische Daten sind im Internet über http://dnb.ddb.de abrufbar.

Deutsche Nationalbibliothek holds a record of this publication in the Deutsche Nationalbibliografie; detailed bibliographical data can be found under http://dnb.d-nb.de

Museum Wiesbaden

HOCHSCHULE FRESENIUS
UNIVERSITY OF APPLIED SCIENCES

1. Auflage 2018

Druck und Bindung
Printing and Binding
De Vega Medien GmbH Augsburg

Printed by deVega

So umweltfreundlich wie möglich produziert. Klimaneutral gedruckt mit mineralölfreien Druckfarben.
Mehr Infos unter www.devega.de

ISBN 978-3-89258-120-8

HISTORISCHE STÄTTE DER CHEMIE
DIESES HAUS ZEUGT ALS LETZTES VON DEM GEBÄUDE-
KOMPLEX, DER DAS CHEMISCHE LABORATORIUM
FRESENIUS WIESBADEN BEHERBERGTE. ES WURDE 1848
VON DEM CHEMIKER, FORSCHER UND LEHRER
CARL REMIGIUS FRESENIUS
(1818 – 1897)
GEGRÜNDET. SEINE FORSCHUNGEN ZUR ANALYTISCHEN
METHODENENTWICKLUNG UND QUALITÄTSSICHERUNG
UND SEINE LEHRBÜCHER FÖRDERTEN DIE
WEITERENTWICKLUNG DER REINEN UND ANGEWANDTEN
ANALYTIK SOWIE DIE INDUSTRIELLE ANWENDUNG
NACHHALTIG. ER PRÄGTE VIELE GRÜNDER-
PERSÖNLICHKEITEN DER CHEMISCHEN INDUSTRIE.
AUS DIESEM LABORATORIUM GINGEN DAS INSTITUT
FRESENIUS IN TAUNUSSTEIN UND DIE HOCHSCHULE
FRESENIUS IN IDSTEIN HERVOR.
ENTHÜLLT AM 18. JULI 2013, IM 100. GEBURTSJAHR SEINES
URENKELS WILHELM FRESENIUS, LANGJÄHRIGER REKTOR
DER HOCHSCHULE FRESENIUS, EHRENMITGLIED DER
GESELLSCHAFT DEUTSCHER CHEMIKER UND
EHRENBÜRGER DER LANDESHAUPTSTADT WIESBADEN.
GDCh GESELLSCHAFT DEUTSCHER CHEMIKER

Gereimtes[43]

Die Ewigkeit

O Mensch! Die Ewigkeit Dir auszudenken,
Vergeblich Müh'n,
Du kannst Dich hoffnungsfreudig nur versenken
In Phantasien.

Denn sieh, auf Erden sind wir eng verbündet
Mit Raum und Zeit,
Drum bleibt uns ewig unerschlossen, unergründet
Die Ewigkeit.

Doch wie sich auch das Jenseits mag gestalten,
Glaub' und hab Mut,
Durch unsres Gottes liebevolles Walten
Wird alles gut!

In rhyme[43]

O Man! In vain you preconceive
Eternity.
In hope and joy, you can but drift
In fantasy.

Lo! On this earth with time and space
Closely allied are we.
Never will be fathomed or revealed
Eternity.

Unknown to us is the hereafter,
But through God's loving governance, this I can tell
– Have faith, take heart! –
All shall end well.

Gedicht von Carl Remigius Fresenius, niedergerschrieben am Vorabend seines Todes. Familienarchiv.

Poem of Carl Remigius Fresenius, written down in the evening before his death. Family archive. (Translated by Leo Gros and John Nicholson)